Are Humans *Really* Responsible for Changing Climate?

LOOKING OUT THE WINDOW

The Trial of Carbon Dioxide in the Court of Public Opinion

Bob Webster

PAGE PUBLISHING, INC.
Conneaut Lake, PA

First originally published by Page Publishing 2021

ISBN 978-1-6624-2920-0 (pbk)
ISBN 978-1-6624-2921-7 (digital)

Printed in the United States of America

CONTENTS

PROLOGUE

This book is intended to help readers examine real-world evidence to better understand whether it supports or contradicts the theorized relationship that claims growing atmospheric CO_2 is causing global climate to warm.

The Birth of the Human-Caused Climate Change Belief

Does Earth face an impending "existential" threat from fossil fuel use?

Many people sincerely believe their future is in real danger of dramatic climate change that will cause severe famine, coastal inundation, unusually severe weather events, and other climate-related catastrophes claimed to be caused by civilization's dependence on fossil fuels.

People who hold this fearful belief generally do so on the basis of dire warnings coming from some climate change scientists. Can these scientists be trusted to provide a full, fair, and honest evaluation of the likelihood of any future climate change threat?

In an ideal world people should be able to rely on those they've entrusted with responsibility to warn them of any future dangers. This book isn't just about evidence and claims that fossil fuels pose an "existential threat" to humanity; it is also about a betrayal of trust.

The human-induced climate change belief is examined in a way that seeks to replace belief with understanding. It is the author's hope that the understanding gained from the perspective and knowledge imparted by *Looking Out the Window* will calm most fears about Earth's future climate and our use of fossil fuel to power modern civilization.

What is meant by climate change?

Climate is generally defined as the *prevalent weather conditions over many years*. It is customary to consider thirty years of temperature change sufficient time to represent a measure of *climate* and multiple years of observed climate a reasonable indicator of *climate change*.

With that in mind, US government records[1] dating from the late 19th century are used to construct graphics for readers to easily visualize any consistent relationship between changing global atmospheric carbon dioxide (CO_2) and changing global average surface temperature (GAST), the key measures associated with climate change theory.

Theory claims that changing atmospheric CO_2 (growing CO_2) is causing global temperatures to change (global warming).

The observed relationship between atmospheric CO_2 growth and changing global surface temperatures is examined in a way designed to be interesting, understandable, and requiring no special scientific knowledge. This examination will clearly reveal the true nature of the relationship established by more than 100 years of global records for atmospheric CO_2 and global average surface temperatures.

Using a 19th-century greenhouse effect theory as midwife, the coincidence of late 20th-century climate warming with a long-term growth of atmospheric CO_2 gave birth to the *greenhouse gas climate change theory*.

The essence of this new greenhouse gas climate change theory asserts that in the absence of a stronger mitigating force:

If atmospheric CO_2 increases, climate will warm.

If atmospheric CO_2 decreases, climate will cool.

Does an examination of the records for atmospheric CO_2 and temperature change offer compelling evidence contradicting this theorized relationship? Should theory prove to be inconsistent with real-world observations, then real world evidence must supersede any theory that inexorably binds changing climate to changing atmospheric CO_2.

The brief coincidence of late 20th-century warming with growing atmospheric CO_2 (Figure 1-1, page 2) is seized upon as evidence supporting this theory while records spanning more than a century that soundly contradict this theory are ignored (Figures 1-2 through 1-4, pages 3 and 4).

The new climate orthodoxy cannot be questioned. Those who ask the obvious questions are treated as heretics and derisively labeled as "skeptics" and "deniers." When records do not support the theory, those who have the temerity to point out that inconvenient truth are slandered and bullied. That isn't science. It's dogma.

A fortuitous late 20th-century solar grand maximum together with a strong El Niño warming spike in 1998 served to produce enough short-term climate warming to build a base from which this new climate change narrative was launched simply because it was accompanied by a persistent long-term atmospheric CO_2 growth.

Naturally, if carbon dioxide growth is to be the culprit for catastrophic global warming, then the source of that growth had to be identified and charged as an accessory.

Despite claims to the contrary, nobody really knows the true source of recent atmospheric CO_2 growth. A convenient patsy was found in the increased global use of fossil fuels with their additional CO_2 emissions into the atmosphere. These emissions became the designated source of atmospheric CO_2 growth. While substantial evidence exists to seriously question fossil fuel as the source of observed atmospheric CO_2 growth, it does make for a good climate change demon if one blindly accepts the validity of the underlying climate change theory.

Ironically, it doesn't seem to have occurred to those promoting this narrative that growing atmospheric CO_2 might simply be a by-product of climate warming that, in turn, warms ocean waters. Warmer oceans will shift the balance of ocean emission and absorption of CO_2 to and from the atmosphere in favor of greater emissions.

Having charged fossil fuel with being responsible for atmospheric carbon dioxide growth, the new narrative quickly found a new label to exploit—anthropogenic (human-caused) global warming (or simply AGW). Blame humanity for using fossil fuels! If this sounds a bit familiar to older generations, during mid-20th century cooling, some scientists blamed fossil fuels for releasing sunlight-blocking particles that were claimed to be causing climate *cooling*.

But along the road to warming Armageddon something unexpected happened. Warming paused. Oops!

After more than a decade of the inconvenient pause in warming made the term *global warming* increasingly awkward, a new label, *human-caused climate change*, quietly emerged. This change facilitated an even broader spectrum of potential disasters for the public to be warned about and, of course, dutifully blamed on the use of fossil fuels. Blame "big carbon" for climate change and the grand human-caused climate-change narrative is complete.

To insinuate the notion that climate wouldn't be changing if not for fossil fuels, the term was revised yet again to just plain *climate change*.

Thus the belief that human activity is responsible for climate change is anchored by two premises:

> Climate *change* (warming) reflects atmospheric CO_2 *change* (growth).

> Atmospheric CO_2 growth is substantially caused by the use of fossil fuels for energy and transportation.

The first premise is based on the somewhat controversial use of the 19th century greenhouse gas theory as the foundation for the related greenhouse gas *climate change* theory characterized above.

Depending on context, throughout this book, this theory will be referred to as *greenhouse gas climate change theory*, *climate change theory*, or just plain *theory*.

Should climate change be found to be entirely independent of the growth of atmospheric CO_2, then the first premise above is invalid and the second premise becomes completely irrelevant.

Rather than explore why atmospheric CO_2 might be growing, this book will focus on global records to examine whether the theorized relationship between atmospheric CO_2 growth and temperature warming is supported by real world evidence (observations in nature).

Looking Out the Window

At some time most readers have probably experienced a day when fair skies and pleasant temperatures were forecast, yet showers and cool temperatures prevailed while the forecast for "fair and pleasant" weather continued.

Had forecasters gone blind? Could they not see the clouds and rain?

In short, forecasters can seem to be so focused on their weather maps and computer models they neglect to *look out the window* to validate their forecast. Have scientists who predict an approaching climate Armageddon failed to verify the basis for the existential threat they perceive? Is there a similarity to weather forecasters with respect to projections of climate change catastrophe?

According to climate history from the mid-18th century to the early 21st century, yes, it is quite evident that proponents of this new greenhouse gas climate change theory either failed to *look out the window* at real-world evidence, or they did look and simply ignored what they saw because it contradicted their theory.

This book is for those who seek a better understanding of the climate change issue from the perspective of actual observed records. Some may be uncomfortable relying on what they've been told to believe, while others may have noticed that many of the claimed consequences of warming simply have not come to pass.

By *looking out the window* at real-world evidence, readers can see for themselves what historic records tell them about the real relationship between changing atmospheric CO_2 and global temperature change.

Consequently, a trial format is loosely adopted to help readers examine the evidence relating to climate change theory. Readers are members of the jury. Defense evidence is presented in the context of a jury trial—specifically, a trial in the court of public opinion.

As the leading advocate for the greenhouse gas climate change theory, the United Nations' Intergovernmental Panel on Climate Change (IPCC) serves as the prosecutor. Except where a specific reference to the IPCC is helpful or more appropriate, any reference to the prosecutor is synonymous with a reference to the IPCC.

To gain a better understanding of this relationship, *the scientific method* will be used to scrutinize climate change theory on the basis of climate records (geologic, ice core, and measured temperature records).

A Trial in the Court of Public Opinion

It is a challenge to examine the relationship between atmospheric CO_2 and climate change for both nonscientists and scientists. This is particularly true when readers have been preconditioned over the course of decades to believe that late 20th- and early 21st-century climate change is substantially and incontrovertibly caused by modern civilization's use of fossil fuels.[2] Doubters of this theory-based belief are dismissed as unqualified to challenge the authority of those "climate change experts" who champion this theory.

The defense simply asks the jury to objectively view real-world evidence from a perspective free of bias and render a fair and just verdict guided by *the scientific method*. Jurors who will benefit most from this examination include many who have little or no scientific background. The average juror is not likely to be familiar with *the scientific method*, therefore, *the scientific method* is defined so jurors will understand why it is the very best process to guide their examination of this trial's evidence.

Before *looking out the window*, readers are introduced to the key players: Earth's atmosphere, carbon dioxide (CO_2), and global average surface temperature. A good perspective on atmospheric CO_2 and climate provide the jury with a sound foundation upon which to build their examination of climate change theory using *the scientific method* applied to the historic evidence relating atmospheric CO_2 change to climate change.

Chapter 1, "Opening Statements for the Defense," outlines the case for the defense. Chapter 2, "In the Beginning…" presents a general overview of the history of the human-caused climate change issue. Chapter 3, "Instructing the Jury: Background Information," introduces the jury to basic information about carbon dioxide and global climate history in straightforward terms with helpful illustrations so the jury can gain an appropriate perspective on these two key players. A look at the 550-million-year geologic record of climate and atmospheric CO_2 is added at the end of Chapter 3 to help jurors apply the perspectives they've gained on the two key players, atmospheric CO_2 change and climate change.

Easy to understand color-keyed graphics are used extensively to help jurors visualize the relationships between key components traditionally associated with climate change.

A glossary is included (after chapters) to explain in very basic terms many of the abbreviations and phrases encountered within this book (e.g., IPCC, *the scientific method*).

The preponderance of defense testimony examines US government-maintained records for observed atmospheric CO_2 change[3-6] and climate change.[7] These records are used to scrutinize the prosecution's theorized relationship between atmospheric CO_2 change and temperature (climate) change.

Key phrases and words are shown in *italics* for special emphasis.

The jury is urged to review the glossary material to understand any new terms or to refresh their memory, a precaution that might prove helpful and lead to a better understanding of the evidence.

Defense exhibits (taking the place of appendices) offer the jury additional supporting evidence not essential to the defense of atmospheric carbon dioxide yet which the jury might find both informative and compelling. The Contents lists topics covered by each defense exhibit.

Our quest for climate change truth begins with this advice from Ephesians 4:14:

> We will not be influenced when people try to trick
> us with lies so clever they sound like the truth.
> *(Holy Bible, New Living Translation)*

If the jury's examination of the relationship between changing atmospheric CO_2 and climate contradicts the prosecution's theory, then that theory must be invalid for not conforming with real-world observations. In that case, the jury must acknowledge that atmospheric CO_2 change *is not a discernible climate change force*, and the source of any atmospheric CO_2 growth is entirely irrelevant, including any portion contributed by fossil fuels or any other human activity.

By *looking out the window* and examining the fragile footing of the foundation upon which rests the prosecution's charges against fossil fuels and atmospheric CO_2, the jury, guided by *the scientific method*, will gain a better perspective from which they can render their most informed verdict in this trial of atmospheric CO_2 in the court of public opinion.

Finally, a few words about why this book was written. Its sole motivation is a principled devotion to truth, in particular, scientific truth. As a mathematician, the author understands that science must be based entirely on scientific truth, not unscrutinized speculative theory or rank conjecture. Emotion has no role in science. Any attempt to shape science to suit a political agenda or social cause, no matter how noble the agenda's proponents believe their cause, is necessarily fraudulent, destructive of science, and by definition, a corruption of *the scientific method*.

When an unchallenged conclusion is mandated as a starting point, truth is sacrificed on the altar of political dogma.

> *Science corrupted by politics is simply*
> *propaganda masquerading as science.*
> —Author, 2019

The public has every right to expect that government agencies do not betray their trust.

Note to readers: Before beginning Chapter 1, readers are encouraged to first take the short multiple-choice quiz found in Defense Exhibit D-1 and record their selections without peeking at the correct answers. After reading Chapters 1–7, retake the quiz and compare your "before" and "after" answers to gauge what you've learned by *Looking Out The Window*.

CHAPTER 1

Opening Statements for the Defense

Atmospheric carbon dioxide (CO_2) is accused of being responsible for climate change during the 20th and early 21st centuries. More specifically, the prosecution alleges that the CO_2 emissions from fossil fuel combustion are substantially responsible for the growth of atmospheric CO_2. That growth is alleged to be the principal cause of global *climate warming*:

> Anthropogenic greenhouse gas emissions have… led to atmospheric concentrations [growth] of carbon dioxide…[that] are extremely likely to have been the dominant cause of the observed warming since the mid-20th century.

> Anthropogenic greenhouse gas (GHG) emissions since the pre-industrial era have driven large increases in the atmospheric concentrations of carbon dioxide… The ocean has absorbed about 30% of the emitted anthropogenic CO_2, causing ocean acidification. (IPCC, *Climate Change 2014 Synthesis Report, Summary for Policymakers*, page 4)[2]

These conclusions are rooted in the prosecution's greenhouse gas climate change theory.

A theory typically relates a precedent action to a consequent condition—for example, "if this, then that." When "this" becomes "*global atmospheric CO_2 changes*" and "that" becomes "*global climate will change accordingly*," the prosecution's greenhouse gas climate change theory can be reduced to:

> If *global atmospheric CO_2 changes*, then *global climate will change accordingly*.

The defense will help the jury examine real-world evidence for consistency with this theorized relationship.

Should the evidence support the prosecution's theory, the defense will concede the prosecution might have a case. However, should the evidence contradict the prosecution's theory, the scientific method invalidates the prosecution's theory and atmospheric CO_2 growth cannot be responsible for global climate warming.

The defense stipulates that relatively short-term transient forces (e.g., massive volcanic eruptions, strong El Niño warming events, strong solar variability) might temporarily overwhelm the theorized relationship between changing atmospheric CO_2 and climate change. Absent any such mitigating force, for the prosecution's theory to be valid, yearly atmospheric CO_2 change should consistently be observed to produce corresponding yearly global average surface temperature (GAST) change. If atmospheric CO_2 concentration goes up, GAST should warm; if atmospheric CO_2 concentration goes down, GAST should cool.

Does the prosecution's theorized cause and effect reflect what happens in the real world? The jury's role is to examine the evidence by "looking out the window" at the recorded relationship between atmospheric CO_2 change and climate change and then assess whether the theorized relationship is consistent with real-world evidence (*theory might be valid*) or inconsistent with real-world evidence (*theory must be invalid*).

When the prosecutor claims a growing concentration of atmospheric CO_2 is causing climate change, it generally shows the jury a graphic such as the 35-year history of atmospheric CO_2 and global average surface temperature (merged land and ocean surfaces) as depicted by Figure 1–1.

As atmospheric CO_2 concentration (red line) grew 59.6 parts per million (ppm) from 1977 through 2011, global average surface temperatures (cyan with red shading) warmed by 0.66°C (1.18°F). Implying such a chart is typical or that it constitutes "proof" that warming temperatures were caused by a growing concentration of atmospheric CO_2 is misleading. While 35 years is sufficient time to represent climate, the prosecution misleads the jury by suggesting the warming is caused by growing atmospheric CO_2.

While Figure 1–1 appears to support the prosecution's theory, Figures 1–2 through 1–4 reveal the prosecution's case is contradicted by the preponderance of the evidence.

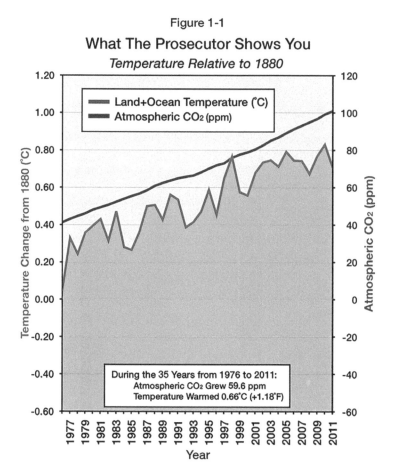

Figure 1-1

What The Prosecutor Shows You

Temperature Relative to 1880

Figure 1–2 shows the relationship between global temperature (cyan with blue shading) and atmospheric CO_2 (red) during the 30 years from 1881 through 1911—again, a sufficient timeframe to represent climate. Note that while atmospheric CO_2 grew by 9.2 ppm, global surface temperatures cooled by -0.37°C (-0.66 °F), a strong contradiction to the prosecution's greenhouse gas climate change theory.

Figure 1-2

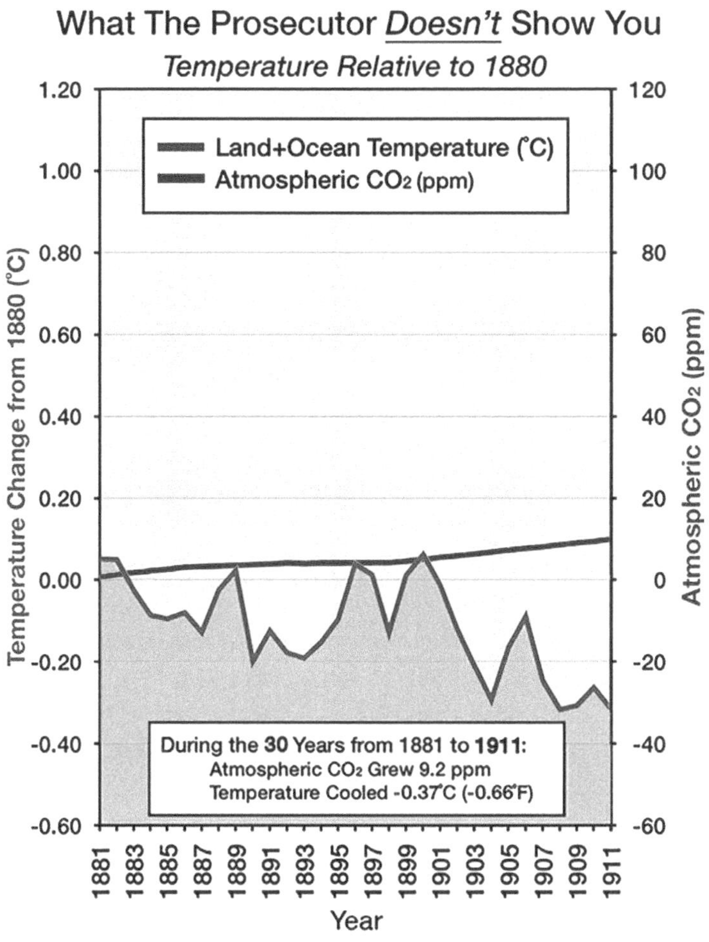

What The Prosecutor *Doesn't* Show You
Temperature Relative to 1880

The historic warming during the 1930s is clearly shown in Figure 1–3, which shows 33 years of climate from 1911 through 1944. As atmospheric CO_2 grew just 9.6 ppm, virtually identical to what it had grown during the preceding 30 years of cooling, global temperatures warmed by an astonishing 0.73°C (1.32°F)!

Figure 1-3

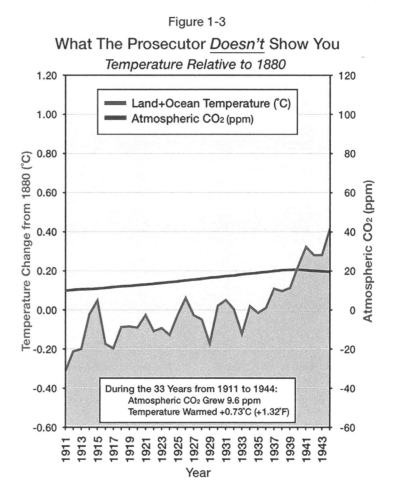

What The Prosecutor *Doesn't* Show You
Temperature Relative to 1880

This represents a theory defying opposite temperature response to virtually the same change in atmospheric CO_2.

Finally, Figure 1–4 shows that during the 32 years from 1944 through 1976 while atmospheric CO_2 was growing by a whopping 21.8 ppm, global temperatures actually *cooled* by -0.38°C (-0.66°F).

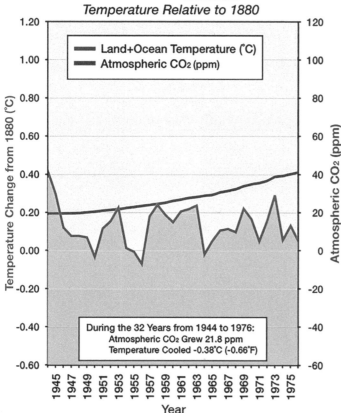

Figure 1-4

What The Prosecutor *Doesn't* Show You

These four distinctly different responses of global climate to changes in the atmospheric concentration of carbon dioxide have been extracted from Figure 4–22 Chapter 4 where 130 years of climate and atmospheric CO_2 change are examined.

How much faith can the jury have in the prosecution's evidence if it dwells on just one portion of a 130-year climate history, the *only* portion that appears to support the prosecution's theory of a crime? Jurors should note, too, that the three portions ignored by the prosecution each dramatically contradict the theory of a crime advanced by the prosecution.

If a comprehensive examination of the evidence were to conclusively show that atmospheric CO_2 change *is not even a detectable* (let alone strong) climate change force, then it doesn't really matter *what* is causing atmospheric CO_2 to grow and *fossil fuels cannot possibly have any impact* on climate change.

The scientific method (described in Chapter 2 and the glossary) provides a straightforward rational process scientists use to discriminate between valid theory and specious theory. This is the method by which the defense will help the jury scrutinize the validity of the prosecution's greenhouse gas *climate change theory* by examining the evidence based primarily on the testimony of global records for atmospheric CO_2 and temperature maintained by US government agencies.

Note that it really isn't necessary to actually examine the science upon which a theory is based. It is only necessary to examine the *evidence in nature* to judge whether the theorized relationship is valid.

A natural question posed by *the scientific method*: Is theory supported by real-world evidence? There are only two possible options. Evidence either supports or contradicts climate change theory. If evidence contradicts theory, then the theorized relationship between atmospheric CO_2 change and climate change is invalid and the prosecution's case collapses. It's really that simple.

When a third option is repeatedly asserted (*that real-world evidence must be faulty*) then theory has become dogma and the issue has moved from the realm of science to the realm of religious doctrine—it must be accepted as a matter of faith, and all evidence to the contrary must be denounced as heresy.

The jury will examine nature's testimony for evidence of a consistent compelling relationship between changing atmospheric CO_2 and surface temperature or climate change.

Does real-world evidence support the prosecution's theorized relationship between atmospheric CO_2 change and global climate change? That question is thoroughly examined in Chapter 4, where US government records for global average atmospheric CO_2 and global average surface temperature (GAST) are scrutinized from the late 19th century through 2019.

The defense raises the following key points for the jury to consider as it scrutinizes the evidence:

- The prosecution's climate change theory claims that in the absence of other strongly mitigating forces, atmospheric CO_2 change is a strong force causing climate change. Consequently, if atmospheric CO_2 persistently grows, climate should warm; if atmospheric CO_2 persistently declines, climate should cool.
- Because the prosecution's theory is based on radiant heat transfer that occurs virtually instantaneously (with the speed of light), any effect from such heat transfer should be measurable without delay, and a full year should provide ample time for the theorized effect to be recorded. Consequently, year-to-year atmospheric CO_2 and temperature change should be consistently theory compliant, particularly over the course of decades spanning more than a century.
- By virtue of the prosecution's repeated assertions that late 20th-century atmospheric CO_2 growth is substantially the cause of observed climate warming, the prosecution has conceded that other strong mitigating forces are absent and that such time spans are sufficient to either support or invalidate its theory.
- It is generally agreed that a minimum of thirty years is required to establish a single instance of climate change evidence. Global CO_2 and temperature records are a basis for creating a clear view of how well thirty-year CO_2 change and thirty-year temperature change support or contradict climate change theory.

The importance of these key points will become more evident to the jury throughout Chapter 4.

The prosecution had already laid out its case and presented its theories to the jury well before members of the jury realized they were part of any trial.

For decades, the jury has been conditioned to believe *there is no defense* ("the science is settled," "the debate is over," "97% of scientists agree…" etc.). Essentially, the prosecution has told the defense to "sit down and shut up!" while the prosecution's agents pile on (news media hyping stories for ratings, lavishly-funded academia that generates supporting "studies," politicians seeking a cause to champion for political advantage, wealthy business interests looking to profit from doomsday fearmongering, etc.).

When speaking of theory, a competent scientist *will never* claim "the science is settled." Only scientific law is considered settled. The very essence of science is to challenge dogmatic orthodoxy and scrutinize unproven theory with a skeptical eye.

Consider that the claim "97% of scientists agree…" is not only faulty; it is a brazen appeal to authority by consensus and a debate-dodging technique used by those who cannot successfully debate their position. Much better to simply shut down any debate.

Consensus *is never* a substitute for honest spirited scientific debate. In science, evidence is compelling; consensus means little.

> *When the debate is lost, slander becomes the tool of the loser.*
> —Socrates

While the jury has heard the terms *skeptic* and *denier* used to brand those who do not subscribe to the prosecution's climate change theory, the jury should be aware that the very essence of science is skepticism. Name-calling is never an acceptable alternative to honest scientific debate.

When a new theory is proposed, a competent scientist will skeptically scrutinize it to challenge its validity. This is not an adversarial process, it is simply a proper use of the scientific method.

The term *climate change denier*, when used to defend a climate change theory, is little more than mindless bullying. It is doubtful there is a competent scientist alive who questions the fact that climate is always changing, though neither the cause nor pace of such change is likely to be readily apparent or universally agreed upon.

The defense stipulates that climate change is real, routine, and frequent throughout Earth's climate history. The jury will examine that history to assess the validity of the prosecution's climate change theory. If the prosecution's theory is valid, then there should be ample evidence in the past to corroborate the theorized relationship between changing atmospheric CO_2 and climate change. Laws of science do not change over time. The science that governs climate change today is no different from that which governed climate change in the past.

On the basis of a relatively small degree of global warming, the prosecution has tried to whip the jury into a frenzy of fear, telling it at every opportunity that it must restrict its use of the most widely available low-cost energy source known, plant-based fossil fuels (coal, oil, and natural gas). Each prosecution witness assures the jury with ever-increasing certainty that catastrophic climate lurks in our near future if swift sweeping reforms are not instituted.

Why? Because "80% of the world's energy comes from fossil fuels"[8] and carbon dioxide (CO_2) gas is emitted into the atmosphere as a byproduct of the combustion of fossil fuels.

The prosecution alleges the relatively trivial amount of CO_2 gas annually released by fossil fuel combustion is rapidly accumulating in the atmosphere and is, in turn, the dominant source of the recent growth of the trace atmospheric gas, CO_2 (~0.04% of the atmosphere's volume). The fossil fuel portion of recent atmospheric CO_2 growth is thus alleged to be responsible for climate change.

The prosecution further alleges that since the early 1940s, unprecedented atmospheric carbon dioxide growth of 40 to 50 ppm (~10% CO_2 growth) amounting to just 0.005% of the atmosphere's volume is causing a rapid increase in global average temperatures.[2]

The prosecution warns that continued use of fossil fuels is likely to cause catastrophic climate warming, rising sea levels, coastal city inundation, ocean acidification, famine, and all manner of weather disasters.

These dire forecasts are predicated on *theorized* climate change consequences catalogued by the prosecution in its indictment of defendant atmospheric carbon dioxide (CO_2).

The fundamental allegation that *recent climate warming must be caused by atmospheric CO_2 growth* will be thoroughly examined by the jury in Chapters 4 and 5.

Controversial theory detail is not explored; instead, the jury will focus on the testimony of real-world evidence to determine whether reality (what really happens in nature) supports or contradicts the prosecution's *theorized* relationship.

Using *the scientific method*, the defense will present compelling evidence to the jury that challenges the prosecution's theorized relationship between changing atmospheric CO_2 and climate change.

US government records from the late 19th century through 2019 will be examined to assess beyond any reasonable doubt, whether or not the prosecution's alleged relationship is supported by *looking out the window* at real-world evidence.

Evidence will be examined over every reasonable timeframe from decades to more than a century of observations and from thousands of years of ice core records to the past half-billion years of geologic evidence.

Since the 2015–2016 strong El Niño (ENSO) is known to have temporarily disrupted normal surface temperature profiles, some of the contemporary evidence is based on records ending with the year 2014, a sufficient timeframe to examine various periods of change.

If the evidence testifies that the theorized relationship between atmospheric CO_2 change and climate change is invalid, then fossil fuels used to power modern civilization can neither be responsible for nor linked to any degree of climate change.

The only testimony defense will offer from unvalidated climate models will demonstrate that the prosecution's climate change models actually testify against the underlying climate change theory upon which they are based (Chapter 5).

To help the jury understand the sequence of this trial, the following chapter outline identifies how each chapter contributes to this trial.

- Chapter 1, "Opening Statements for the Defense," describes how the jury will scrutinize the prosecution's theorized relationship between atmospheric CO_2 and climate change.
- Chapter 2, "In the Beginning…" describes the origin of the climate change issue. *The scientific method* is defined, and the defense case is outlined.
- Chapter 3, "Instructing the Jury—Background Information," provides jurors with the required perspectives on (1) defendant carbon dioxide and (2) Earth's climate history. The relationship between climate and atmospheric CO_2 as revealed by the 550-million-year geologic record is examined.
- Chapter 4, "Testimony of Global Average Surface Temperature and Atmospheric CO_2," examines the US government's records for annual global average atmospheric CO_2 and global average surface temperature (GAST) to scrutinize the relationship between annual global atmospheric CO_2 change and annual GAST change. Does changing atmospheric CO_2 concentration cause corresponding change in surface temperature and climate? That key question is addressed by examining the testimony of various long-term records that should provide clear evidence that either supports or contradicts the prosecution's climate change theory.
- Chapter 5, "Who are the *Real* Climate Deniers? More Defense Testimony," presents compelling evidence that bears directly on this case. Much of this evidence focuses

on observations since the early 1980s when the phrase "human-caused global warming" emerged on the heels of nearly four decades during which scientists were deeply concerned about a cooling Earth and the possible dawn of a new ice age! Clear evidence the prosecution's climate simulation models *strongly testify against* the prosecution's climate change theory is presented.

- Chapter 6, "Summation of Carbon Dioxide's Defense," summarizes carbon dioxide's powerful defense to the prosecution's theory-based evidence-challenged allegation that climate change since the mid-20th century is a direct consequence of the growth of atmospheric carbon dioxide.
- Chapter 7, "Defense's Closing Statement to the Jury," offers jurors the defense's closing remarks. Jurors are encouraged to independently come to their own conclusions guided by *the scientific method* and the evidence presented in Chapters 1 through 6 and powerfully supported by Defense Exhibits A through D.

Following are brief descriptions of defense exhibits:

- Defense Exhibit A: "Disinformation in the news and reliable information sources." Cites typical climate change news reports and deconstructs misleading claims. These are examples of how news stories have misrepresented the issue and why so many people have been misled to accept the prosecution's climate change narrative without questioning its veracity. Contains a sample of the many good independent resources for jurors who wish to learn more about this issue from an unbiased source.
- Defense Exhibit B: Some additional evidence helpful to the defense case, including ocean acidification and the new USCRN records in some detail from 2005 through 2019. USCRN records are from the 114 most well-maintained

and accurate new recording stations carefully sited throughout the contiguous 48 US states, none of which suffer from any corrupting local development issues.

- Defense Exhibit C: A partial list of the key evidence brought out in testimony for the defense of atmospheric CO_2 that might be helpful during the jury's deliberation.
- Defense Exhibit D: Six theory-challenging exhibits, (D-1) a climate change quiz, (D-2) a short counterexample to the climate change narrative, and (D-3 through D-5) three compelling pieces of evidence covering 2005–2019, 1912–1976, and 1880–2018, and finally, (D-6) an analogy to put atmospheric CO_2 into a realistic perspective.

Summary: Opening Statements for the Defense

1. The jury will examine evidence relating to the fundamental charge in the prosecution's indictment to determine whether it is little more than *rank speculation* based on an *unproven specious climate change theory*.
2. Using *the scientific method* together with global atmospheric CO_2 and climate records, the defense will present compelling evidence that defendant is innocent of all charges.
3. Evidence will be presented to show:

   a. Contrary to the prosecution's theory, no relationship beyond chance exists between changes in atmospheric CO_2 concentration and changes in global average surface temperatures.

   b. No evidence of causation exists between *either* atmospheric CO_2 and global average surface temperature *or* atmospheric CO_2 *change* and global average surface temperature *change*.

4. The prosecution's theory-based climate change studies and reports cannot alter the scientific fact that if records prove atmospheric CO_2 change and temperature (climate) change are *uncorrelated,* atmospheric CO_2 change *cannot possibly be the source* of global atmospheric climate change as alleged by the prosecution.

5. The defense will present compelling evidence that *total* atmospheric CO_2 has no discernible impact on global climate over any time period examined during the past 550 million years of Earth's climate history. If atmospheric CO_2 has no impact on climate, then the fraction of atmospheric CO_2 attributable to fossil fuel emissions *cannot possibly impact climate.*

6. The defense is convinced that a fair jury, upon examining compelling real-world evidence, will come to the only possible conclusion:

> *No causal relationship exists between atmospheric CO_2 change and climate change in any record over any meaningful period of time. Because changing atmospheric carbon dioxide has no impact on climate change, the source of any atmospheric carbon dioxide growth is entirely irrelevant, and fossil fuel use has no impact on climate change.*

CHAPTER 2

In the Beginning...

The Scientific Method

The scientific method is a method scientists use to test *the validity of a theory or hypothesis* by one or more of the following:

a.   Evaluating its conformity with scientific laws (*Is any scientific law violated?*)
b.   Testing by experimentation (*Is it compatible with experimentation?*)
c.   Real-world observation to assess conformity with reality (*Does it conform with the observed evidence in nature?*)

If a theory or hypothesis *fails any one of these tests*, it is invalid and *must be* rejected or revised.

To have the greatest confidence in a theory, it should survive scrutiny of all three tests of *the scientific method*. However, failure of *any one of them* is sufficient to invalidate any theory.

When addressing a theory's validity, neither the complexity of the theory's scientific derivation nor the number of scientists (consensus) who endorse the theory are relevant factors. If a theory is contradicted by any one of the tests of *the scientific method*, the theory is, by definition, *invalid*. Consequently, *the scientific method* is a valuable tool to help scientists invalidate specious theory.

On the other hand, *the scientific method* can lend support to a theory if (1) the theory *does not violate* any known scientific principles, (2) the theory *conforms with* experimentation, and (3) the theory *is consistent with* real-world observations.

Note that *the scientific method* cannot validate a theory, it can only fail to invalidate a theory.

Jurors should understand *the scientific method* because (1) it is a straightforward process by which a fair and competent scrutiny of any theory can be performed, and (2) *the scientific method* will be used to guide the jury throughout this trial.

Noted Scientists on *the Scientific Method:*

> Albert Einstein: "No amount of experimentation can ever prove me right, a single experiment can prove me wrong."

> Nobel Laureate Richard Feynman: "It doesn't matter how beautiful your theory is, it doesn't matter how smart you are. If it doesn't agree with experiment, it's wrong."

Paraphrasing Feynman:

> It doesn't matter how beautiful your theory is, it doesn't matter how smart you are. If it doesn't agree with observation, it's wrong.

Does human-caused climate change theory rest upon a valid foundation?

Fortunately, *the scientific method* provides the jury with a straightforward rational process by which the validity of the greenhouse gas climate change theory can be examined. Using *the scientific method*, the path to invalidate a specious theory (one that seems valid but isn't) is remarkably clear of obstacles. If real-world evidence (observation) contradicts theory, there is no need to understand or debate the complexities of the theory or any supporting hypotheses and assumptions upon which the theory may depend. Nonconformance with real-world observation will *invalidate any theory* regardless of how fervently its supporters believe it to be valid.

For example, the human-caused climate change theory implicates fossil fuel emissions as the source of atmospheric carbon dioxide (CO_2) growth that is claimed to be responsible for climate change. If the theorized relationship between atmospheric CO_2 growth and climate warming is contradicted by the preponderance of real-world observations, then the theorized relationship must be invalid, the prosecution's case evaporates, and the source of atmospheric CO_2 growth becomes entirely irrelevant.

When scrutinizing any theory, a natural question of an inquisitive mind should be, does real-world evidence support or contradict the theory? Those are the only two options.

Should evidence invalidate the prosecution's theory, any assertion that fossil fuel emissions are responsible for the growth of atmospheric CO_2 is irrelevant. Should the evidence be inconclusive, then the theory is not validated; it can only be said to have not been invalidated.

A more powerful tool to examine the theorized relationship between atmospheric CO_2 change and temperature change is their correlation.

Correlation Coefficient

A statistical tool known as *correlation coefficient* (range: -1.0 to 1.0) indicates whether a causal relationship is possible between two measures (does an *action* lead to a *reaction*?). A negative correlation (less than zero) means a positive action causes a negative response (or vice versa). Using records corresponding to the theorized relationship (changing atmospheric CO_2 leads to a changing climate), a correlation coefficient can be calculated to see if a causative relationship exists.

If the two records are correlated, then there *might be* a causal relationship between the two (i.e., growing atmospheric CO_2 *might cause* climate warming). If the the recorded measures are uncorrelated, there *cannot be any causal relationship* between the two. The existence of correlation does not guarantee a cause-effect relationship, but *the lack of correlation guarantees there can be no cause-effect relationship.*

Correlation is described in greater detail in the glossary.

A correlation coefficient (r) in the range 0.7 to 1.0 is an indicator that two data sets are positively correlated and an action-reaction relationship *might exist* between both measures. If two data sets are uncorrelated ($0 \leq r < 0.7$), a causal relationship between the two is impossible.

The Human-Caused Climate Change Narrative

After more than a decade of sharp climate warming during the "dust bowl" years of the 1930s, NOAA records show Earth's climate went through nearly forty years of post-dust bowl climate cooling, hitting bottom in the late 1970s.

Ironically and contrary to the prosecution's theory, for many of those cooling years, atmospheric CO_2 was *dramatically growing* at a substantially greater rate than had been observed during the dust bowl years of strong warming prior to 1940. This alone constitutes significant evidence that something is very wrong with the climate change theory.

Records show atmospheric CO_2 began to sharply increase every year since 1944. This dramatic growth of atmospheric CO_2 fueled speculation that it was causing unusual global warming (now climate change). But what do NOAA's records[1] actually show?

Referring to Table 2-1, examine the period 1912 through 1976. NOAA and NCDC records show that during the period 1912 through 1944, global average atmospheric CO_2 grew by just nine ppm while GAST sharply warmed by 0.55°C (0.99°F). Yet completely contrary to the

prosecution's climate change theory, during the period 1945 through 1976, records show global average atmospheric CO_2 grew dramatically by 22 ppm as GAST cooled by -0.28°C (-0.50°F).

Table 2-1
Clear Evidence Contradicting Climate Change Theory
(1912 through 1976)

Time Period	CO_2 Change	Temperature Change	Yearly CO_2 Change	Yearly Temperature Change
1912 through 1944	+9 ppm	+0.55°C (+0.99°F)	+0.28 ppm/year	+0.017°C (+0.031°F)
1945 through 1976	+22 ppm	-0.28°C (-0.50°F)	+0.71 ppm/year	-0.009°C (-0.016°F)

Source: NOAA, NCDC

Summarizing these records, when atmospheric CO_2 grew modestly (at 0.28 ppm/year), Earth's climate sharply warmed 0.99°F (0.031°F/year). Yet, when atmospheric CO_2 dramatically grew 22 ppm, 246% (0.71 ppm/year) more than it had when Earth's climate sharply warmed, the climate dramatically cooled -0.50°F (-0.016°F/year).

When atmospheric CO_2 grew modestly, climate sharply warmed. Yet, when atmospheric CO_2 grew sharply, climate cooled! This clear example shows jurors that for more than sixty years during the 20th century, the prosecution's theory was defied by the clear evidence contained in the records for atmospheric CO_2 and temperature maintained by NOAA and NCDC.

Because thirty years is sufficient time to represent evidence of climate change, this example constitutes clear evidence that, for its failure to agree with the evidence found in nature, something is very wrong with the prosecution's greenhouse gas (GHG) climate change theory.

During the latter phase of the pronounced cooling period (1945–1976), some scientists were concerned about the possibility that climate cooling signaled the beginning of the end to the 10,700-year-old

Holocene interglacial and warned of the possible beginning of a new ice age cycle.[9]

Other scientists believed the cooling was human-caused. They thought fossil fuel burning had created atmospheric particles that were blocking sunlight, thereby cooling Earth's surface, just as for several years the massive 1815 Tambora and 1883 Krakatoa volcanic eruptions had cooled global climate. ("The year 1816 is known as 'The Year Without a Summer' in New England because six inches of snow fell in June and every month of the year had a hard frost."[10])

Cooling concerns mounted until about 1980 when a remarkably sudden transition occurred. In a relatively short time span concurrent with the approaching peak solar activity of the modern solar maximum (Figure 3–4, page 43), cooling ceased and post-Little Ice Age[11] warming resumed.

The end of global cooling caused an abrupt reversal by many of the same scientists who had warned of a coming ice age! What hasn't changed is that humans are still being blamed, but this time for "unprecedented" and potentially "catastrophic" global warming because they use fossil fuel for energy!

Worth noting is that diminishing Arctic Sea ice extent has been a favorite graphic used to claim polar bears are threatened by fossil fuels that are warming the planet. Yet Arctic Sea ice in the late 1970s was at its greatest extent since before the early 20th-century warming! The nearly four decades of cooling had expanded year-round sea ice cover in the Arctic to its 20th-century peak in the late 1970s.

Many comparisons of sea ice extent during the final decades of the 20th century and into the 21st century are biased by the ice extent that had peaked in the 1970s and has since naturally and slowly diminished over the ensuing decades.

That sudden warming coincident with the persistent growth of atmospheric CO_2 motivated a small group of scientists to take action.

A controversial theory commonly known as the "greenhouse effect" was resurrected from its 19th-century obscurity, dusted off, recast by the prosecution as the "greenhouse gas climate change theory," and used to boldly proclaim humanity's use of fossil fuel is creating rapid atmospheric growth of a trace gas, CO_2 that, in turn, is causing "unprecedented" global warming that will become "catastrophic" if drastic action to reduce "carbon emissions" isn't taken without delay.

Thus the grand climate change narrative emerged based on a theory of greenhouse gas driven climate change. This narrative, based entirely on an unvalidated climate change theory, has repeatedly been cited in schools and academia and by politicians and news media as the basis for claiming global warming is human-caused and a modern crisis. Yet it isn't scientific law; it remains just an unproven theory.

The prosecution's indictment of atmospheric CO_2 is rooted in IPCC pronouncements [*emphasis* added]:

> Anthropogenic greenhouse gas emissions have… led to *atmospheric concentrations of carbon dioxide*…that are unprecedented in at least the last 800,000 years…and *are extremely likely to have been the dominant cause of the observed warming since the mid-20th century.*

> Anthropogenic greenhouse gas (GHG) emissions since the pre-industrial era have driven large increases in the atmospheric concentrations of carbon dioxide (CO_2)…[40% of] cumulative anthropogenic CO_2 emissions…have remained in the atmosphere… The ocean has absorbed about

30% of the emitted anthropogenic CO_2, causing ocean acidification. (IPCC, *Climate Change 2014 Synthesis Report, Summary for Policymakers*, page 4)[2]

Defense testimony will scrutinize every aspect of those statements, though it will focus on records for both global average atmospheric CO_2 and global average surface temperature (GAST) anomalies to examine the real-world evidence of the relationship between atmospheric CO_2 change and observed temperature and climate change.

Either this relationship will conform with the prosecution's theorized relationship or, like the example summarized in Table 2-1, it will fail to conform, in which case the prosecution's theory (and the associated narrative) will be invalidated for failure to be consistent with observation as required by the scientific method.

While it is easy to criticize the prosecution for creating climate-change fear on the basis of unvalidated theory-based allegations lacking supporting evidence, the prosecution's approach is understandable when viewed in context with the IPCC charter[12] that directs it:

> [To] assess on a comprehensive, objective, open and transparent basis the scientific, technical and socio-economic information relevant to understanding *the scientific basis of risk of human-induced climate change*, its potential impacts and options for adaptation and mitigation. [*emphasis* added]

In fact, the IPCC is directed by its charter to assume climate change is "human-induced" and to do everything in its power to warn the populace of consequences based on that assumption!

While the IPCC may be forgiven for blindly following its charter, there is no excuse for independent scientists blindly accepting the IPCC's theory-based allegations against defendant carbon dioxide without seriously scrutinizing the IPCC's theory for conformity with real-world evidence.

Defense testimony will not rely on unproven theory. Key testimony in Chapter 4 is drawn from *the evidence* found in established historic global records maintained by the US government documenting a global history of atmospheric CO_2 and global average surface temperature anomalies.

Should real-world evidence contradict the dictates of unproven theory, then *the scientific method demands* the evidence *must prevail* over theory. Jurors are not free to ignore the compelling testimony of real-world evidence, regardless of how often the prosecution waves in the jury's face what its theory-based "studies say" or "reports indicate."

Additional testimony in Chapter 5 and Defense Exhibit B (*Additional Topics Supporting Defense Testimony*) challenge key prosecution allegations by drawing on an established scientific knowledge base of subject matter experts together with evidence assembled from government records and both government and nongovernment scientists.[13]

Unaware of the existence of compelling evidence to the contrary, pervasive repetition of the climate change narrative has reached the point where many, including some scientists, are convinced it is a valid theory.

Persistent news stories warning of an approaching climate Armageddon have led to a belief that even if we do not know whether the theory is valid, we must follow the precautionary principle[14] that assumes the worst and take the necessary steps to eliminate modern civilization's dependence on abundant, relatively inexpensive fossil fuels.

> A strong conviction that something must be done is the parent of many bad measures. *(Daniel Webster)*[15]

As Daniel Webster warned, this ill-conceived principle has created demands that humanity convert to green renewable alternatives to fossil fuels, regardless of the cost or the evidence supporting any basis for rash, premature action (e.g., *The Green New Deal*).

Another example of a rash emotional response is that taken by the editorial board of *USA Today*, who opined they were sufficiently satisfied a looming climate disaster is real on the basis *of a single hot July in 2018*. The editorial they penned *"a hellish July validates climate change forecasts"*[15] is pure sophistry. Of course, the hot July in 2018 proved nothing, just as the cold winter in 2019 proved nothing. Short-term weather patterns vary enormously and are not necessarily related to changing global climate. Except, apparently, for the *USA Today* editorial board, who cherry-pick stories and use them for fearmongering fodder designed to increase profits by boosting circulation.

Yet another emotional response is the *Paris Climate Agreement* and its proposal to limit fossil fuel emissions, a costly folly in the absence of any serious real-world evidence supporting the claim that CO_2 growth is a strong climate change force. Such purely emotional motivation also produced a bizarre proposal in the US to "green" modern civilization in the vision of the *Green New Deal*. The underlying belief is that the enormous cost to save Earth from catastrophic global climate change *must be endured* to avoid the *existential threat* climate change theory promises. Costly rash action in the absence of a shred of real-world evidence supporting the climate change narrative and its underlying theory isn't prudent, it's foolish!

Just as the weatherman who forecasts a sunny day while refusing to look out the window at the pouring rain, climate alarmists on the *USA Today* editorial board have never bothered to examine whether real-world evidence supports the theorized relationship between growing atmospheric CO_2 and Earth's climate. While individual claims of record-high temperatures were frequently heard during July 2018, not one of the fifty US states' all-time high temperature records were set in 2018. Not one. Yet one state all-time low temperature record (-38°F, Illinois) was set in January 2019. The jury heard nothing from *USA Today's* editorial writers about that record.

Mesmerized by claims of looming disaster, political leaders across the globe hastily propose solutions to a problem whose existence they haven't yet confirmed by "looking out the window" at the evidence. This is akin to a criminal prosecutor indicting someone in the absence of any real evidence of a crime and then focusing on the punishment without bothering with the inconvenience of a trial.

This jury will examine the testimony of observed evidence relating changing atmospheric CO_2 to climate change as revealed by geologic and ice core evidence. It will also examine contemporary (1880–2019) records that reveal the true relationship between observed global average atmospheric CO_2 and global average surface temperature (GAST).

Though closely related, the prosecution's *greenhouse gas climate change theory* (changing atmospheric CO_2 causes climate change) *should not be confused with the greenhouse effect* that refers to the reaction of certain atmospheric gas molecules to thermal energy Earth's surface radiates at various limited wavelengths of the infrared spectrum. The greenhouse effect can only impact the rate at which the Earth-atmosphere system cools (it cannot raise Earth's surface temperature). At current atmospheric concentrations, additional quantities of greenhouse gases have little (if any) detectable impact on Earth's global average surface temperature (see Defense Exhibit B). Earth's atmosphere moderates its surface temperature. Lacking an atmosphere, Earth's surface would be hotter during the day and colder at night, just as it is on Earth's Moon.

Without some guidance, most of the jury is unlikely to have the scientific background required to scrutinize the science related to the prosecution's climate change theory. It is for this reason that the defense will facilitate the jury's use of *the scientific method* to determine whether real-world evidence *supports* or *invalidates* the prosecution's climate change theory.

The jury should understand that only *scientific laws* are considered "settled science"; everything else is unsettled theory. Conscientious scientists subject theory to skeptical scrutiny. (Einstein's *theory of general relativity* continues to undergo skeptical scrutiny that has led to its refinement and improvement.)

Theory must conform with observation, i.e., what happens in nature. Any theory that does not must either be refined until it does or be rejected as hopelessly invalid.

Looking Out the Window at the Evidence

Perhaps it's time for a serious *look out the window* at the evidence found in Earth's atmospheric CO_2 and climate histories.

If real-world evidence (observation) supports climate change theory, then appropriate steps can be taken to try to identify the source of atmospheric CO_2 growth. In that circumstance, if humanity's use of fossil fuels is confirmed to be the dominant source of atmospheric CO_2 growth, then appropriate steps could be taken to assure the public any such significant human influence on climate change is adequately mitigated.

On the other hand, if *looking out the window* at real-world evidence reveals the claimed significant relationship between changing atmospheric greenhouse gases and climate change does not exist, then the amount of carbon dioxide in the atmosphere is entirely *irrelevant to climate change* and its source of growth is immaterial.

This book is written specifically for those who prefer to *look out the window* at real-world evidence to examine climate change claims before jumping to conclusions and embarking on costly solutions to mitigate a problem that may not exist (e.g., the *Paris Climate Agreement*, the *Green New Deal*).

By *looking out the window*, this jury will use *the scientific method* to rationally scrutinize the prosecution's theory. To that end, well-established historic annual records maintained by the US government for global average atmospheric CO_2 and global average surface temperature will be examined.

Belief in unproven theory must be suspended in favor of a serious examination of the evidence. This is particularly important when for decades many credible scientists have noted key failures of the predicted consequences of climate change theory and raised legitimate objections to the questionable science underlying the greenhouse gas climate change theory.

As most scientists will agree that changing climate is an ongoing process, the key questions to be addressed become:

1. Is recent observed climate change significant, unnatural, unusual, or unprecedented?
2. Does any evidence exist to confirm a relationship that changing atmospheric CO_2 causes climate change?
3. Is recent observed atmospheric CO_2 growth really caused by modern civilization's use of fossil fuels?

Reliable answers cannot be found by reviewing theory or the evidence from theory-based climate models or any studies or reports based on the premises that underlie climate change theory. Such material inevitably operates on the same premises as the theory being scrutinized. These questions are more appropriately addressed by examining the evidence found in nature by *looking out the window* at the records.

All that is required of jurors is a patient examination of the records for time periods sufficient to determine whether evidence consistently supports or contradicts claims arising from greenhouse gas climate change theory.

Readers looking for a detailed examination of climate change theory should look elsewhere. This book is for those who want an honest assessment of the evidence Earth's climate is changing significantly and, if so, whether that evidence confirms the observed change is strongly related to atmospheric CO_2 change as the prosecution claims and its climate change theory predicts.

If climate change should consistently show no more than a chance relationship with atmospheric CO_2 change, then the greenhouse gas climate change theory will be invalidated by virtue of its nonconformance with observation as required by *the scientific method*.

Other crimes of carbon dioxide have been alleged, including the claim that ocean acidification is caused when CO_2 emissions build up excessively and are absorbed by oceans that create acids, causing ocean water to become more acidic. This ancillary claim is refuted by expert testimony in Defense Exhibit B.

While this trial should be strictly a matter of honest scientific investigation and debate, unfortunately it has taken on a highly political nature. A political agenda is a very poor rationale for promoting regulations based on unvalidated scientific theory.

The prosecution and its agents have had exclusive dominion over this trial and have freely attacked defendants atmospheric CO_2 and fossil fuel CO_2 emissions while the defense has not had equal standing to present its evidence or cross-examine the prosecution's evidence.

As Sir Winston Churchill once observed, "A lie gets halfway around the world before the truth has a chance to get its pants on."

It is common knowledge that the prosecution routinely bullies, insults, and blockades the defense to the point the public barely realizes there is a defense.

Members of the defense team who object and speak out (including highly-qualified scientists, politicians, public figures, and university faculty) expose themselves to ridicule, loss of professional standing, bullying, and even job loss.

As if that weren't enough, physical threats from climate bullies supporting the prosecution's dogma are not uncommon risks for those who dare question the climate change dogma being preached by the prosecution and its agents.

Infused with political considerations from the outset, after lopsided debate losses with skeptics in many public forums, the prosecution changed strategy and now shuns debate entirely, claiming "the science is settled" and "the debate is over." How convenient.

This situation has led a number of misinformed politicians to weigh in on a subject about which they have no proper perspective, little or no education, and scant personal or professional knowledge or background. When challenged, such politicians typically appeal to the authority of the prosecution (the IPCC).

Is Public Opinion Well-Informed?

Hardly a week goes by without at least one reference in the news to climate change (*shorthand for human-caused climate change*):

> "The Arctic seems to be warming up. Reports from fisherman [and] seal hunters…who sail the seas [of] the eastern Arctic, all point to a radical change in climate conditions, and…unheard-of high temperatures…"

> "The hot dry seasons of the past few years have caused rapid disintegration of glaciers in Glacier National park, Montana… Sperry Glacier…has lost one-quarter or perhaps one-third of its ice in the past 18 years… If this rapid rate should continue…the glacier would almost disappear in another 25 years…"

> "All the glaciers in Eastern Greenland are rapidly melting, declared [a] Swedish geologist… 'It may without exaggeration be said that the glaciers, like those in Norway, face the possibility of a catastrophic collapse.'"

> "Polar temperatures are on an average six degrees higher than those registered…40 years ago. Ice measurements are on an average only 6-1/2 feet against 9-1/2 to 13 feet…"

> "The ice of the Arctic Ocean is melting so rapidly that more than one-third of it has disappeared in fifty years…"

> "An Arctic expert said…Polar icecaps were melting at an astonishing and unexplained rate and were threatening to swamp seaports by raising the ocean levels… 'The glaciers of Norway and Alaska are only half the size they were 50 years ago.'"

> "Born about 4,000 years ago, the glaciers that are the chief attraction in Glacier National Park are shrinking so rapidly that a person who visited them ten or fifteen years ago would hardly recognize them today as the same ice masses."

The jury is told fossil fuels (coal, oil, natural gas) are responsible for recent global warming and warned their continued use could bring "catastrophic climate change" that poses an "existential" threat.

The possibility that climate change might be natural is virtually never heard. Dire warnings designed to raise fears about the dangerous consequences of climate change (rising seas, more violent storms, more frequent hurricanes, etc.) motivate rash, hastily-conceived political solutions that are unlikely to have any discernible impact on future climate. The obvious purpose of such hyperbole is to motivate humanity to precipitously change its primary energy sources to those that do not create CO_2 emissions. Is this a noble course of action or pure folly?

Would it surprise jurors to learn that the above-quoted dramatic news stories warning of dangerous climate warming were originally published in 1922, 1923, 1939, 1940, 1947, and 1952 (twice), respectively? It appears that climate change hysteria is nothing new.

It might interest jurors to learn that during 2019, National Park Service employees began quietly removing "Gone by 2020" signs from park displays after glaciers resumed growing circa 2010. For some reason, that news never made headlines in either the evening cable news or the morning newspapers.

Worth rereading is the second quote from 1923 and the last quote from 29 years later in 1952, four years after the 1923 quote claimed Sperry Glacier "would almost disappear"! Jurors reading the 1952 quote might believe Sperry Glacier is just a distant memory in 2021 (nearly 100 years after the dire 1923 warning).

For those concerned about the current status of Sperry Glacier, the following quotation is from the United States Geological Survey (USGS) website:

> Sperry Glacier occupies a broad, shallow cirque situated just beneath and west of the Continental Divide in the Lewis Mountain Range of Glacier National Park, Montana. This northeast facing glacier is wider than it is long relative to its flow direction and spans about 300 m in elevation with a median altitude of 2450 m. *It ranks as a moderately sized glacier for this region…*[17] [*emphasis added*]

What key points should the jury take away from Sperry Glacier's story?

1. During a period of similar climate warming, atmospheric CO_2 growth in the early 20th century was radically less than it was in the late 20th and early 21st centuries.
2. Glacier National Park glaciers formed only 4000 years ago(!) as the Holocene interglacial began to cool.
3. Climate alarmists have a lot in common with *Chicken Little*.†

_____

† Chicken Little, *noun*: one who warns of or predicts calamity especially without justification—*Merriam-Webster dictionary*

Against a strong current of climate disinformation, the voices of caution are drowned out (a selection of similar poorly-informed contemporary climate change news stories are deconstructed in Defense Exhibit A).

The condition of the Arctic ice cap is frequently cited as evidence of climate change (meaning human-induced climate change caused by fossil fuels).

To put Arctic ice change in perspective, consider that the open water Northwest Passage route across northern Canada was first sailed by Norwegian Roald Amundsen more than a century ago.

More than half a century later, *during one of the coldest phases* of a nearly four-decade mid-century cooling trend, the US submarine *Skate* surfaced in an open pool of water at the north pole (August 11, 1958).[18] Clearly, there are localized forces (changing ocean currents, undersea volcanoes and other geologic thermal activity, etc.) that are capable of causing Arctic Sea ice to melt, and yet such forces are virtually ignored by the prosecution because, as they are completely unrelated to global atmospheric carbon dioxide, they don't fit the prosecution's climate change narrative.

Imagine the headlines blaming "global warming" and "climate change" if a submarine were to surface in open water at the North Pole during late summer next year. Despite such an occurrence having a precedent more than sixty years earlier, the collective roar from climate alarmists would be deafening. Proponents of the prosecution's climate change theory would demand the United States rejoin the *Paris Climate Agreement*. There would be renewed demands for mandatory carbon footprint restrictions, carbon taxes to fight climate change, and renewed pressure to adopt *the Green New Deal*. Yet there is no evidence open water at the North Pole is related in any way to atmospheric CO_2 growth. None whatsoever! The prosecution does not base its claim of human causation on any real evidence; it's merely *theorized* without being *proven*.

Following the early 1980s birth of the global warming hysteria, a number of rational thinkers including scientists, scholars, journalists, and other public figures questioned the human-caused global warming theory.[19–22] Rather than welcoming scrutiny and engaging in serious debate, the prosecution dogmatically defends its theory. This is the reaction despite the theory's many flaws that are easily exposed by simply *looking out the window* at the evidence.

Some jurors might be inclined to accept the climate change orthodoxy without question. After all, the average juror isn't likely to possess the scientific background to confidently question peer pressure that blindly accepts the prosecution's climate change narrative so frequently and fervently parroted in schools, publications, and by nearly every news outlet.

Without a clear understanding of Earth's long historic relationship between global climate and atmospheric CO_2, jurors will lack the requisite perspective to comfortably question the merits of the prosecution's theory.

Jurors who are uncomfortable with how little they understand natural climate change should take heart in the fact that even the best scientists have yet to develop a validated and generally-accepted understanding of climate change science.

Why? Earth's climate is both extremely complex and affected by a host of earthbound forces in conjunction with a number of complex extra-planetary solar and orbital forces[23–27] that combine to make discovering and validating climate science both difficult and elusive. Toss in the occasional significant impact from space and there will always be some uncertainty about future climate change because some factors are simply not predictable.

Powerful disparate forces work to alter global climate. The truth is, nobody understands climate change sufficiently to accurately project the future course of climate. An accurate, verified, and validated computer simulation model for Earth's climate change simply does not exist. Climate projection models are of little value if their underlying theory is invalid. To make its case, the prosecution relies upon its theory-based models' evidence while ignoring a host of theory-contradicting evidence found by *looking out the window*.

The evidence found in records for global average atmospheric CO_2 and temperature can be examined for any support of the prosecution's claim that atmospheric carbon dioxide growth is a strong force causing significant and potentially catastrophic global climate warming.

The prosecution's reliance on theory-based climate simulation models and a vague reference to warming climate in preference to a serious examination of the evidence contained in global records for atmospheric CO_2, temperature, and climate suggests the prosecution isn't entirely convinced of either the veracity or validity of its own theory.

No matter how sophisticated the coding or honorable the intentions, no matter how competent the developers, the brutal reality is that the prosecution's CMIP5 climate simulation models built upon an invalid theory will always over-predict temperature and are thus doomed to failure. The best way to hide that failure is to shun any examination of the real-world evidence contained in records for global atmospheric CO_2 and temperature.

Forces controlling global climate change are vastly more complex and likely more plausible than the prosecution's greenhouse gas climate change theory. The jury should never accept the prosecution's theory-based *assertions* as *facts*.

The geologic evidence from the most recent 550 million years, the GISP2 Greenland and 800,000-year Vostok Station Antarctic ice core evidence, and the recorded evidence dating from the late 19th century are all capable of testifying to the real nature of any relationship between changing global atmospheric CO_2 and climate change. Yet the prosecution appears uninterested in thoroughly examining that testimony.

On a variety of time scales, climate change can be strongly influenced by unpredictable natural processes ranging from solar variability to geologic upheavals and major impacts of objects from space. There is no nice, tidy theory that can predict future climate. There never will be.

It is no wonder scientists have yet to reliably identify those factors having the greatest influence on Earth's climate and climate change. There are many theories but very little certainty when it comes to predicting future climate—lots of smoke and mirrors, very little fire or clarity. Yet one thing is abundantly clear, the complex science of climate change remains very much elusive and *unsettled*. Claims to the contrary are motivated either by ignorance or by a dishonesty necessitated by a political agenda that is out of harmony with the real world.

Just as in its past and without any human influence, Earth's future climate will naturally include both bitter cold and steamy heat. At some point, rather than dwell on the false claims of an invalid theory, civilization will do well to prepare for inevitable dramatic *natural* climate change episodes that cannot be prevented.

Outline of the Case for the Defense

In particular, the defense of atmospheric CO_2 will focus on the testimony of four key witnesses:

1. US government records for global average atmospheric CO_2
2. US government records for global average surface temperature (reported as anomalies)
3. The 800,000-year Antarctic (Vostok Station) and the nearly 11,000-year Greenland (GISP2) ice core records for both atmospheric CO_2 and climate (temperature)
4. Earth's 550-million-year geologic record for both atmospheric CO_2 and climate (temperature)

If real-world evidence *consistently* and *unequivocally contradicts* the prosecution's greenhouse gas climate change theory, then *the scientific method will have invalidated* that theory just as it would invalidate any other theory found to be inconsistent with real-world evidence.

When jurors discover what geologists have learned from Earth's engrained geologic evidence of past climate, past climate change, and past atmospheric CO_2 and combine that discovery with more recent records showing the contemporary relationship between climate change and atmospheric CO_2 growth, the jury will have ample evidence to confidently assess the validity of the prosecution's claimed relationship between changing atmospheric CO_2 and global climate change.

Jurors are encouraged to read through each chapter in sequence. Following the key evidence in Chapters 4 and 5, jurors might want to explore additional evidence in Defense Exhibits A through D before moving on to Chapters 6 and 7.

Further independent investigation is encouraged and can be pursued by exploring the many references provided and online sources listed in Defense Exhibit A. Most references include a link to online source material. Links sometimes change. If not redirected to a new link, then perform a search on the reference's key phrases (e.g., the *italicized* reference title).

Before continuing on to Chapter 3, to improve familiarity with some important terms used throughout this trial, jurors may want to briefly review the glossary of terms and abbreviations (following the Epilogue).

After *looking out the window*, the jury is strongly encouraged to *reach its own verdict* on the basis of the evidence presented. To that end, the jury will need to set aside any prior beliefs and biases acquired from exposure to either unsubstantiated material acquired from repetitive news reports and peer group preconceptions or from the pronouncements of politicians and academia whose views are molded by theory-based "expert" assertions accepted without proof, cross-examination, or thorough scrutiny of the evidence found by *looking out the window*.

The testimony of real-world observations must be reflected in the jury's verdict. No matter how enticing it may be, theory that fails to agree with evidence is deeply flawed, invalidated, and should not be given any credibility.

It is commonly known that while a long list of domestic and international government and nongovernmental agencies and organizations supportive of the prevailing climate change narrative can be compiled, virtually *every one* of those endorsements *exclusively relies* on the acceptance of theory-based claims made by the prosecution (UN's IPCC). Consequently, to the extent any consensus exists, it is *a consensus to rely on the claims of a single source*, the IPCC (prosecution). Such reliance is nothing more than a consensus of one!

Jurors should give particular emphasis to Chapter 4's testimony by the evidence contained in government records for global atmospheric CO_2 and global average surface temperature anomalies from the late 19th century to the present.

The primary charge in the prosecution's indictment of defendant carbon dioxide:

- Recent climate warming is principally caused by atmospheric CO_2 growth.

Ancillary charges levied by the prosecution:

- Fossil fuel CO_2 remains in the atmosphere for 50 to 200 years or more.
- Atmospheric CO_2 growth is substantially caused by fossil fuel combustion.

Should the principal charge be invalid, both ancillary charges are entirely irrelevant. Chapters 3–5 will provide jurors with the evidence they need to confidently proceed to a verdict on all charges.

Summary: *In the Beginning…*

1. The scientific method is a rational process by which any theory or hypothesis can be scrutinized for validity. Every theory must agree with the established laws of science (scientific laws), experimentation to test the theory (experimentation), and the observed evidence in nature (observation).
2. Correlation between two measurements means the two *might be* causally related (a change of one produces a corre-

sponding change in the other). The degree of correlation is measured by the correlation coefficient between two records.

3. If a causal relationship exists between atmospheric CO_2 growth and climate warming (or atmospheric CO_2 and temperature), there must exist a strong positive correlation between the two. While two records might be highly correlated, they might still not be causally related. However, causation cannot exist without correlation.

4. The defense will help the jury examine the records for atmospheric CO_2 and global average surface temperature to determine whether the evidence supports the prosecution's claim that the two are causally related (changing atmospheric CO_2 produces corresponding global average surface temperature change).

CHAPTER 3

Instructing the Jury—Background Information

For jurors to render a proper verdict they must have a good perspective on Earth's atmosphere, carbon dioxide's nature, and Earth's global climate history. A realistic perspective prepares the jury for their examination of Earth's records documenting atmospheric carbon dioxide and global average surface temperature. This examination will scrutinize the veracity of climate change theory,[3-6] and with *the scientific method* as their guide, will enable the jury to reach a just verdict.

Earth's Atmosphere

Without its atmosphere, Earth's surface would be much like its moon's, barren of life, much hotter during daylight hours and much colder at night. Clearly Earth's atmosphere serves to *moderate* its surface temperature by making days *cooler* and nights *warmer* than they would be without an atmosphere. Earth's weather and climate are defined by its atmosphere.

Nearly all of the Earth's atmosphere is made up of only five gases: nitrogen, oxygen, water vapor, argon, and carbon dioxide. Several other compounds are also present. Although this…table does not list water vapor, air can contain [slightly more than 4%] water vapor, more commonly ranging from 1–3%. …water vapor [is] the third most common gas…[28]

A generally-accepted figure for the mean atmospheric concentration of water vapor is 1.5% of atmospheric gases. That figure represents the typical water vapor concentration at Earth's global average surface temperature[29] of 14.5°C (58.1°F). Based on this value, atmospheric gases at sea level are ranked in Table 3-1 by their concentration in Earth's atmosphere. The four non-trace (~1% or more) gases accounting for 99.96% of Earth's atmosphere are listed with a light yellow background.

Table 3-1
Atmospheric Composition
(Adusted for mean of 1.5% water vapor)

	Normalized for *1.5% Water Vapor	Normalized Atmospheric Portion (ppm)
Nitrogen (N2)	76.9%	769242
Oxygen (O2)	20.6%	206349
* Water Vapor (H2O)	1.5%	15000
Argon (Ar)	0.9%	9201
Carbon Dioxide (CO2)	trace	404
Neon (Ne)	trace	17.9
Helium (He)	trace	5.2
Methane (CH4)	trace	1.7
Krypton (Kr)	trace	1.1
Hydrogen (H2)	trace	0.5
Nitrous Oxide (N2O)	trace	0.5

* Mean value of 1.5% (based on GAST of 14.5°C) used (actual value varies between a trace and 4% of all atmosperic gases).

Source: Modified* from *Atmospheric Chemistry*, Wikipedia, (https://en.wikipedia.org/wiki/Atmospheric_chemistry).

This trial is about the theorized role of *carbon dioxide* as a climate change force. To a lesser degree, it is also about whether or not the principal source of Earth's recent atmospheric CO_2 growth is the atmospheric accumulation of CO_2 emissions produced by fossil fuel combustion.

The jury should bear in mind that water vapor is generally thirty-seven times more abundant in the atmosphere and reacts with (absorbs) Earth's radiant heat (IR) over a spectrum so broad that it dwarfs carbon dioxide's reactive spectrum (see Defense Exhibit B, Figure B-1).

Carbon Dioxide (CO₂)

Carbon dioxide is a gas molecule consisting of one atom of carbon (C) bonded with two atoms of oxygen (O_2). CO_2 molecules exist in two states:

1. Commonly, as a colorless, odorless, invisible gas
2. Less common, as a solid (frozen, "dry ice")

CO_2 is not found naturally in a liquid state.

The gas bubbles in seltzer, soft drinks, beer, and champagne are CO_2.

CO_2 is principally derived *naturally* from animal respiration, plant decay, the combustion (oxidation) of carbon compounds (brush fires, forest fires) and ocean outgassing.

Anthropogenic CO_2 is derived from human action, primarily from fossil fuel combustion and, to a much lesser extent, non-hydraulic cement production.

Regardless of its source, all CO_2 behaves identically.

Both plant and animal life are carbon-based. Without carbon and the CO_2 gas molecule there would be no plant or animal life on Earth.

Plants grow by absorbing carbon from their environment through photosynthesis. During photosynthesis, plants absorb CO_2, extracting the carbon while expelling the oxygen (O_2) in a symbiotic relationship with animals who breathe in oxygen for metabolism and expel CO_2 through respiration. Carbon dioxide is, indeed, the *gas of life*, without which *there would be no life on Earth*. Given CO_2's critical role in sustaining plant and animal life, it is grossly inappropriate to label atmospheric CO_2 a pollutant.

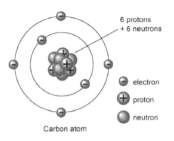

6 protons
+ 6 neutrons

− electron

+ proton

neutron

Carbon atom

The concentration of CO_2 in Earth's atmosphere is usually denoted in parts per million (ppm or ppmv) of atmospheric volume (the average number of CO_2 molecules in each million molecules of Earth's atmospheric gases).

As noted above, CO_2 is absolutely essential for all plant and animal life on Earth. In order to sustain life, Earth's atmospheric volume of CO_2 *must exceed* 135 ppm or 0.000135 (0.0135%) of Earth's atmospheric volume. The current volume of atmospheric CO_2 (about 410 ppm) is just 0.000410 (0.041%) of total atmospheric volume. Put another way, 99.959% of all atmospheric gases *are not* carbon dioxide!

A Quick Overview of Global Atmosphere and CO_2

Earth's current level of atmospheric CO_2 *is extremely low; it is neither high nor unprecedented.*

The prosecution claims fossil fuel emissions are responsible for about half the growth of atmospheric CO_2 during the 20th century (about 37 ppm). To arrive at its figure, the prosecution claims fossil fuel CO_2 emissions have an implausibly high atmospheric residency time, ten to forty times longer than longstanding independent studies show. Chemists, chemical engineers, and other scientists specializing in atmospheric CO_2 contend[31-33] the real residency time of CO_2 emissions is at most four to six years, contributing just 3 to 4 ppm to the recent growth of Earth's atmospheric CO_2, not 37 ppm.

To help understand the scarcity of CO_2 in the atmosphere, consider the following visualization. The Rose Bowl stadium in Pasadena, California has 92,542 seats. Imagine the seats in that stadium are white and they represent all atmospheric gases.

Suppose 38 of those 92,542 stadium seats are chosen at random and 34 of them are painted green while the remaining four are painted orange. The 34 green seats represent every molecule of natural atmospheric carbon dioxide while the four orange seats represent the prosecution's very high estimate of fossil fuel's contribution to 20th century carbon dioxide growth. For perspective, chemical analyses show the actual numbers should be less than half an orange seat with the remainder being green seats.

Suppose jurors were cruising over the empty Rose Bowl stadium in a hot air balloon. Looking down, jurors might reasonably think every seat was white. How likely is it that any of the four seats that are orange would be noticed? Would any of the 34 green seats be noticed? Does this visualization help jurors perceive the relative scarcity of carbon dioxide molecules in the atmosphere?

Jurors should bear in mind two key facts not commonly known: (1) *most of Earth's surface-radiated heat (infrared, IR) is not at wavelengths where CO_2 is most reactive* (*every seat* would look white to those wavelengths); and, (2) the contribution to atmospheric warming declines with each incremental addition of CO_2 to the atmosphere. The diminishing impact on atmospheric temperature as atmospheric CO_2 grows is not in dispute. At current levels of atmospheric CO_2, additional

CO_2 emissions have little, if any, discernible impact on Earth's global climate, a condition strongly supported by real-world evidence.

The prosecution's theory defies any sense of proportion or logic; rather, it resembles a grand hoax built upon a highly dubious, yet specious theory. See Defense Exhibit B for more in-depth discussion of CO_2.

Considering one of Earth's coldest ice eras occurred when atmospheric CO_2 averaged *more than ten times the current atmospheric concentration* (see Table 3-2, page 47), the prosecution's theorized impact of atmospheric CO_2 growth on climate change is highly suspect.

Modern plant life evolved during the carboniferous period between 360 million years ago and 290 million years ago when Earth's volume of atmospheric CO_2 reached up to at least three times the current atmospheric volume of CO_2 (or about 1200-1400 ppm). Earth's atmospheric CO_2 volume *over the past half-billion years* has *averaged more than five times higher* than at present. Any claim current levels of atmospheric CO_2 are "unprecedented" is clearly untrue (see Table 3-2) and misleading.

In contrast, water vapor (H_2O), the most dominant greenhouse gas, varies more substantially, measuring between 10,000 ppm and 20,000 ppm (1% and 2%) at sea level (and varying as high as 42,400 ppm or 4.2% at 86°F). Table 3-1 (page 26) assigns a prevalent 1.5% to water vapor and adjusts other gases accordingly. The atmospheric volume of water vapor is dependent on atmospheric temperature—a colder atmosphere holds less water vapor, a warmer one holds more.

Water vapor (H_2O) reacts to radiant heat from Earth's surface (long-wave infrared, IR) over a vastly greater range of the infrared spectrum than does any other greenhouse gas including CO_2 (see Figure B-1, Defense Exhibit B). To the extent such warming exists, H_2O's far greater abundance and much broader IR reactive spectrum will dominate any greenhouse gas effect generated by trace gases that are more IR reactive but *whose reactivity is constrained to relatively little of the radiant IR spectrum emitted by Earth's surface.*

The *carbon cycle* is an ongoing natural exchange of carbon between Earth's atmosphere (as CO_2) and its biosphere, in particular its oceans. Oceans, covering 71% of Earth's surface, dominate Earth's surface area and are the greatest contributor to the annual carbon cycle.[34]

Oceans have a virtually limitless capacity to absorb fossil fuel CO_2 emissions without significantly affecting ocean pH (acidification).[34] Colder ocean waters absorb relatively more CO_2 from the atmosphere; warmer ocean waters emit relatively more CO_2 to the atmosphere. Therefore, CO_2's absorption from and emission to ocean waters is temperature-dependent. Ocean plant life (phytoplankton) plays a significant role[35] in the seasonal emission and reabsorption of CO_2 to and from oceans.

> Microscopic marine phytoplankton play a critical role in regulating today's carbon cycles, yet not enough is known about the process. These tiny organisms consume carbon dioxide from the atmosphere and move it to marine sediments in the deep ocean in a process called "the biological pump." Currently, more than 99 per cent of the Earth's carbon is bound up in these sediments, locked away in the depths of the ocean.[35]

Given oceans' dominance of Earth's surface area, the life cycle of phytoplankton could represent a significant contribution to the annual carbon cycle, particularly in the southern hemisphere where oceans cover 81% of the surface with 29% of the remaining 19% of surface being glacier-covered year-round (Antarctica), where no plant life contributes to the carbon cycle.

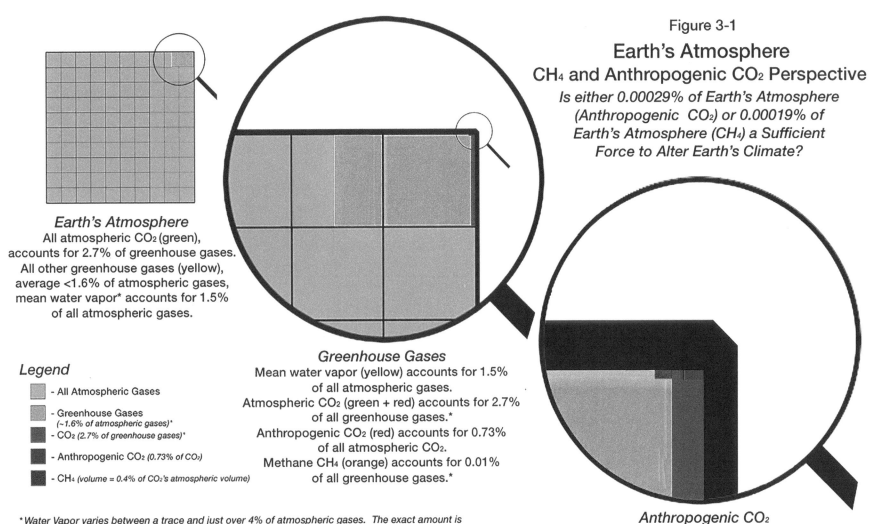

Figure 3-1

Earth's Atmosphere
CH₄ and Anthropogenic CO₂ Perspective

Is either 0.00029% of Earth's Atmosphere (Anthropogenic CO₂) or 0.00019% of Earth's Atmosphere (CH₄) a Sufficient Force to Alter Earth's Climate?

Earth's Atmosphere
All atmospheric CO₂ (green),
accounts for 2.7% of greenhouse gases.
All other greenhouse gases (yellow),
average <1.6% of atmospheric gases,
mean water vapor* accounts for 1.5%
of all atmospheric gases.

Legend

- All Atmospheric Gases

- Greenhouse Gases
  *(~1.6% of atmospheric gases)**

- CO₂ (2.7% of greenhouse gases)*

- Anthropogenic CO₂ (0.73% of CO₂)

- CH₄ (volume = 0.4% of CO₂'s atmospheric volume)

Greenhouse Gases
Mean water vapor (yellow) accounts for 1.5%
of all atmospheric gases.
Atmospheric CO₂ (green + red) accounts for 2.7%
of all greenhouse gases.*
Anthropogenic CO₂ (red) accounts for 0.73%
of all atmospheric CO₂.
Methane CH₄ (orange) accounts for 0.01%
of all greenhouse gases.*

Anthropogenic CO₂
Anthropogenic CO₂ (red) accounts for 0.73%
of all atmospheric CO₂ (green + red).
Fossil Fuel CO₂ (red, to right of vertical black line)
accounts for 87% of anthropogenic CO₂.

*Water Vapor varies between a trace and just over 4% of atmospheric gases. The exact amount is
dependent upon temperature (colder air holds less water vapor, warmer air holds more). For the
purpose of this illustration, a mean value of 1.5% is used based on a global average temperature
of 14.5°C, other atmospheric gases were adjusted accordingly (gases are generally listed in the order
of their atmospheric concentration, ignoring water vapor). Atmospheric Methane (CH₄) from all sources.

Source: Modified* from *Atmospheric Chemistry*, Wikipedia, (https://en.wikipedia.org/wiki/Atmospheric_chemistry).

Figure 3-1 puts into perspective the relative concentration of all atmospheric gases and greenhouse gases, in particular, CO_2, CH_4 (methane), and that portion of CO_2 attributable to human activity (anthropogenic). Bear in mind that atmospheric concentrations are just part of the potential to impact global climate. Except for H_2O, the potential reactivity of greenhouse gases to IR is constrained by the very limited portion of the IR spectrum over which they are reactive to Earth's outbound radiant (IR) thermal energy (see Figure B-1, Defense Exhibit B); water vapor *dominates* atmospheric absorption for most IR wavelengths over which both CO_2 and CH_4 are reactive.

The small 10×10 grid at the upper left of Figure 3-1 represents Earth's atmosphere, each grid representing 1% of atmospheric volume. The two light yellow squares at the upper right of the grid represent all atmospheric greenhouse gases (slightly more than 1.5% of the atmosphere). Dominated by water vapor, the cream-colored squares also include trace gases (methane and carbon dioxide) that, unlike H_2O's reactivity over most of the IR spectrum, are reactive to IR over *extremely limited* portions of the IR spectrum, meaning only a small portion of IR from Earth interacts with CO_2 and CH_4.

The middle view magnifies the greenhouse gas portion of the atmosphere to reveal the portion of greenhouse gases attributable to atmospheric carbon dioxide (CO_2) (in green, just 2.7% of *all greenhouse gases*).

Based on longstanding chemical analyses,[31–32] the lower right magnification shows the human-produced (anthropogenic) portion (in red) of atmospheric CO_2 (about 3 ppm or 0.73%), and the vertical black line divides the portion of anthropogenic CO_2 according to its origin as either from fossil fuel combustion (87% of anthropogenic CO_2, to the *right* of the vertical line) or from all other anthropogenic sources (13%, to the *left* of the vertical line). Other sources include land use and cement production. The jury can dismiss any notion of fossil fuel

CO_2 being a strong climate change force if total atmospheric CO_2 growth is found not to be causally linked to climate change.

Even if the jury were to accept the prosecution's dubious claim that ~42% of CO_2 growth is attributable to fossil fuel use, then between 1940 and 2014, fossil fuel emissions must have accumulated 37 ppm in the atmosphere.

In comparison with atmospheric CO_2 (in green and red), atmospheric CH_4 (methane, in orange) is trivial (2 ppm versus 410 ppm for CO_2). Cow flatulence is a minor source of atmospheric CH_4. The impact of atmospheric methane on either atmospheric temperature or global climate is immeasurably small owing to its reactivity being constrained to just a few extremely limited portions of the IR spectrum where IR absorption is already dominated by water vapor (see Figure B-1, Defense Exhibit B).

Recall the Rose Bowl stadium visualization of the scarcity of fossil fuel-derived atmospheric CO_2. Methane is represented by *less than one seat in six* Rose Bowl stadiums totaling 555,252 seats. Furthermore, only one of the three small IR spectra to which methane is reactive overlaps appreciable IR wavelengths Earth's surface radiates, and IR in that portion of the spectrum is already fully absorbed by atmospheric water vapor. Cow flatulence (methane) as a climate change force is a farce.

Summary: Carbon Dioxide (CO_2)

1. CO_2 commonly exists as a gas and rarely as a solid (dry ice). It does not exist in nature as a liquid.
2. CO_2 can be either *organic* or *inorganic* in origin.
3. Natural origins include animal respiration, plant and animal decay, brush and forest fires, volcanic eruptions and ocean outgassing. Anthropogenic origins include fossil fuel combustion and non-hydraulic cement production.

4. Oceans dominate annual absorption and emission of atmospheric CO_2. Oceans are easily capable of absorbing all the CO_2 that fossil fuel combustion can produce without any meaningful impact on ocean acidification.[34, 36] Cold oceans readily absorb CO_2; warm oceans readily emit CO_2.[37]

5. CO_2 is not a pollutant. On the contrary, *it is essential* for all life on Earth; should atmospheric CO_2 fall below 135 ppm, all life on Earth would become extinct. Labelling CO_2 a pollutant is both ignorant and inappropriate. CO_2 is the gas of life.

6. Today's atmospheric CO_2 levels are significantly below (~1/6th) *the average* for the past 550 million years of Earth's climate. Many contemporary plant and animal species evolved over this same 550 million years when the concentration of atmospheric CO_2 was at least twice what it is today.

7. Compared with atmospheric CO_2 levels when plants evolved, Earth's current atmosphere is CO_2-starved.

8. While methane (CH_4) is highly reactive to IR, it is reactive over a negligible range of the IR spectrum. Both of the small spectra where methane is reactive are outside the main spectrum of Earth's radiant IR and both are overwhelmed by H_2O's IR absorption. Because it is so rare in both quantity and spectral coverage, methane has no measurable impact on global climate or atmospheric temperature. Bovine flatulence, a fraction of atmospheric CH_4, is not a climate change force; it is a climate change farce.

Climate Perspective

Few jurors are likely to possess a perspective on Earth's climate history to appreciate the changing nature of global climate and its natural variability. The following quick overview will put Earth's recent global climate into a perspective that should allow jurors to better understand how current climate relates to the long history of Earth's climate.

A Quick Overview of Global Climate

Figure 3-2 (next page) puts current climate into perspective with global climate over the past 3.5 billion years (the period after single cell life first appeared on Earth). The 10×10 grid at the upper left represents 3.5 billion years of Earth's global climate. Each grid represents 1% of the past 3.5 billion years or 35 million years. The orange-red squares represent 90% of Earth's climate during the past 3.5 billion years when climate was typically about 10°C (18°F) warmer than during the depth of an ice era and 7.5°C (13.5°F) warmer than any climate humans have known![38] This grid shows characteristic climate types, it is not chronological.

The blue squares, totaling 350 million years, represent the sum total of seven ice eras Earth's climate has endured during which global average temperatures drop as much as 10°C (18°F) below Earth's typical climate.

Earth's current climate is experiencing the most recent of those seven ice eras where temperatures range dramatically colder than those characteristic of Earth's much warmer typical climate.

The middle magnification shows the darker blue (nearly two) squares representing the current ice era that began about 65 million years ago and continues today. The white area is the (Pleistocene) ice epoch that began 2.4 million years ago and continues today. Note the much darker blue near the bottom of the white ice epoch area. That very dark blue area shown in the lower right magnification represents the most recent ice age/interglacial cycle that began 115,000 years ago.

Below the dark blue and above the cell border is a very thin yellow line (barely detectable) representing the Holocene interglacial that began 10,700 years ago. Imperceptibly small, a red band at the right side of the yellow interglacial represents the most recent 170 years since the end of the Little Ice Age cool period.

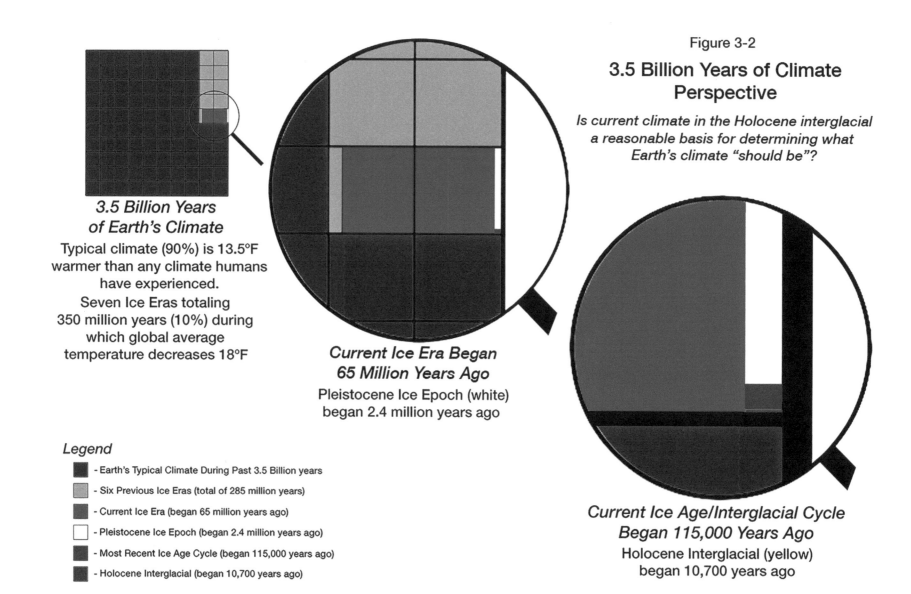

3.5 Billion Years of Earth's Climate

Typical climate (90%) is 13.5°F warmer than any climate humans have experienced.
Seven Ice Eras totaling 350 million years (10%) during which global average temperature decreases 18°F

Figure 3-2

3.5 Billion Years of Climate Perspective

Is current climate in the Holocene interglacial a reasonable basis for determining what Earth's climate "should be"?

Current Ice Era Began 65 Million Years Ago

Pleistocene Ice Epoch (white) began 2.4 million years ago

Current Ice Age/Interglacial Cycle Began 115,000 Years Ago

Holocene Interglacial (yellow) began 10,700 years ago

Legend

- Earth's Typical Climate During Past 3.5 Billion years
- Six Previous Ice Eras (total of 285 million years)
- Current Ice Era (began 65 million years ago)
- Pleistocene Ice Epoch (began 2.4 million years ago)
- Most Recent Ice Age Cycle (began 115,000 years ago)
- Holocene Interglacial (began 10,700 years ago)

Modern humans appeared about 5,000 years ago (about half the yellow band) and developed the earliest civilizations. The oldest human-recorded temperature records (in central England) go back just 300 years. The jury need only compare the overall grid (representing 3.5 billion years of climate) with the lower right magnification for a good perspective on the insignificance of current climate in the grand scheme of Earth's global climate history. Looking at Figure 3-2, is it at all reasonable to expect the past 70 years of the 10,700-year Holocene interglacial (an imperceptible 0.65%) to be a legitimate basis upon which to assert what climate *should be?*

For nearly forty years, the persistent human-caused climate change narrative has created the impression that if not for humans using fossil fuel, 20th-century climate would represent what "should be" or what is normal for Earth.

A more realistic perspective tells an entirely different story. From *Ice Ages* (*Planet Earth* book series):

> Human beings have never experienced the Earth's normal climate. For most of its 4.6-billion-year existence, the planet has been inhospitably hot or dry and utterly devoid of glacial ice. Only seven times have major ice eras, averaging roughly 50 million years in length, introduced relatively cooler temperatures; humankind arose during the most recent of those periods.[38]

"Human beings have never experienced the Earth's normal climate." That's something for the jury to reflect upon.

Recent climate history is archived in government-maintained records of global average surface temperature (GAST) over ocean, land, and merged land plus ocean surfaces. Expressed in terms of temperature and atmospheric CO_2, Antarctic ice core records reveal 800,000 years of climate history. Greenland ice core records detail the Holocene (current) interglacial (10,700 years and counting), and a climate history going back billions of years is recorded in Earth's geologic evidence.[7, 39]

Since climate *change* is this trial's concern (if CO_2 changes, does climate change accordingly?), jurors need an accurate perspective from which to assess the theorized relationship in context with Earth's long history of *natural climate change*. Without an understanding of historic climate change during the past 3.5 billion years, the jury will lack the required perspective to render an informed verdict in this trial. Armed with a good perspective on Earth's long history of climate and climate change, jurors will be better prepared to scrutinize contemporary records.

Earth's Changing Appearance

To appreciate the influence of the changing shape and distribution of continents and oceans (geologic change) on climate change, jurors should be aware of the magnitude and frequency of both significant climate change *and* dramatic surface feature changes[26] during the past 3.5 billion years.

The most elementary cellular life is estimated to have first appeared about 3.5 billion years ago, making the geologic history of Earth's past 3.5 billion years of climate and climate change worth examining. To relate to that time scale, Earth's 3.5-billion-year climate history is scaled proportionally to a 3,000-mile journey along Interstate 80 (I-80) as it crosses the United States. This 3,000-mile journey begins at the New York end of the George Washington Bridge between New York and New Jersey several miles east of the beginning of I-80. The end of this journey through climate history is, today, at the terminus of I-80 in San Francisco, California.

As an example of this perspective, the past 5,000 years of modern human existence is represented by slightly more than the final 22 feet of the 3,000-mile journey to San Francisco. For comparison, 22 feet is a typical front to rear depth of a residential garage in the US.

Geologic change alters the size, shape, and distribution of Earth's continents and oceans as they would appear from space. Such changes are an important aspect of long-term climate change. Relative to a human lifespan, the constant change of Earth's continents and oceans is virtually imperceptible. The movement of Earth's continental plates means that every 50 to 100 million years, Earth's appearance is radically altered.[26] During the jury's hypothetical 3,000-mile (3.5-billion-year) journey, dramatic changes to Earth's landscape would typically occur every 64.3 miles along the journey (nearly 47 times), or roughly *every 75 million years*. We live on a geologically turbulent planet; Earth's familiar appearance from space did not exist 75 million years ago and will be gone 75 million years from now. Radical geologic changes can significantly affect Earth's ocean currents and climate. Two squares in Figure 3-2 represent 70 million years, a period of change during which Earth's appearance from space would become unrecognizable.

Earth's Typical Climate during the Past 3.5 Billion Years

As noted earlier, Earth's biosphere began to harbor carbon-based life about 3.5 billion years ago. Since that time, Earth's climate has experienced a number of relatively brief but dramatic changes. For *90% of that time*, Earth's climate has been relatively stable, averaging 7.5°C (13.5°F) *warmer* than any climate modern humans have experienced.[38] There are *no* year-round polar ice caps at sea level during Earth's typically much warmer climate.

In perspective with the 3,000-mile climate journey along I-80, Earth's typically much warmer climate would be encountered along a total of 2,700 miles (90%) of that journey. The dinosaur era spanned roughly 200 million years (5.7%) of Earth's typical climate, ending as the current ice era began. Relative to the 3000-mile trip across the US, the dinosaur era covered just 171.4 miles.

Ice Eras

Totaling 350 million years (300 miles of the cross-country trip), seven ice eras have interrupted Earth's typically very warm climate. Each interruption averaged just 43 miles of the journey (50 million years). Earth's current climate is embedded within an ice epoch of an ice era and is much colder than Earth's typically very warm climate.[38]

While ice eras are much colder than Earth's typical climate, it is only during ice epochs (the very coldest climate regimes of ice eras) that continental glaciers expand from the poles toward the equator. Six of those seven ice eras have occurred within the past billion years, suggesting Earth's climate is trending *much colder* than what has been typical for most of the past 3.5 billion years.

Earth's climate is currently experiencing one of those rare ice eras. Having begun 60 million years ago, the current ice era is already 20% longer than an average ice era. If Earth were not currently in an ice epoch of an ice era, during their respective summer seasons, both Greenland and Antarctica would be completely ice-free. Yet if climate warmed precipitously to Earth's typical climate, alarmists would be demanding costly restraints on CO_2 emissions, actions that would have no measurable impact on climate because atmospheric CO_2 neither triggers nor ends an ice era. It would be foolish to ignore theory-contradictory evidence and then compound that foolishness by relying on the contradicted theory to guide policy.

Scientists do not know for certain what triggers either the beginning or the end of an ice era. What they *do know* is that during the past 60 million years, Earth's climate has been *significantly colder* than Earth's typical climate. There is speculation and some evidence that link the beginning of the current ice era with the impact of an asteroid or comet the size of a city in the area of the present-day Mexican Yucatán Peninsula; however, the near coincidence of these two events could just as well have occurred by chance. Scientists simply do not know.

Having begun 60 million years ago, in terms of our hypothetical trip across the USA, the current ice era began just 51 miles *from the end* of the 3,000-mile journey, or just two-thirds the distance from Sacramento to San Francisco!

When leaving Sacramento on the final leg of our cross-country trip along I-80, Earth is experiencing its typical climate with the temperature about 13.5°F warmer than any climate humans have experienced. Then, about 25 miles along I-80 toward San Francisco (about one-third of the way), the climate turns sharply colder with the start of the current ice era, eventually plunging temperatures as much as 10°C (18°F) during ice epochs. For the remainder of the trip there would be moderate periods alternating with bitter cold ice epoch climate, though overall climate would still remain much colder than Earth's typical much warmer climate.

Earth has been in this extremely cold atypical ice era climate for the last 51 miles (60 million years) of the 3,000-mile journey, and yet the prosecution claims "unprecedented" warming of a few degrees will doom Earth! If nothing else, the prosecution has certainly mastered hyperbole.

Considering the 7.5°C (13.5°F) warming needed to match Earth's typically very warm climate, does the prosecution's claim of "unprecedented" climate change strike the jury as being dramatically inconsistent with the known history of Earth's natural climate? The post-Little Ice Age impact of recent climate warming cannot reasonably be viewed as either catastrophic or unprecedented nor does it give reason to believe human activity has played any discernible role.

At some point, the current ice era will end and Earth will return to its typically much warmer climate, and it will do so naturally without warning or any help from humans, just as it has six other times over the past 3.5 billion years.

Ice Epochs

As mentioned above, bitter cold climate periods during ice eras are known as *ice epochs*. Earth is currently in the *Pleistocene* ice epoch that began 2.4 million years ago (the final 2 miles of the 3,000-mile journey along I-80). It is the sixth ice epoch of the current ice era.[38] An ice epoch's duration is typically between hundreds of thousands of years up to several million years.

Ice epochs are characterized by a series of ice age-interglacial cycles of varying lengths. Less severe ice age cycles tend to be shorter (~40,000 years), while the coldest climate regimes on Earth have occurred during the longest ice age cycles lasting roughly 100,000 or more years.[40–41] (see Figure 5-22). Even during ice epochs of ice eras, it is only during the longest (coldest) ice age/interglacial cycles that year-round polar ice exists.

Ice Ages and Interglacials

Earth's coldest climate regimes occur during *ice age cycles* within ice epochs of ice eras. During longer ice age cycles, continental glaciers advance from year-round frozen poles toward the equator as Earth experiences its coldest global climate.

In the early Pleistocene ice epoch, ice age-interglacial cycles were characteristically shorter with no year-round polar ice. In the most recent half of the ongoing Pleistocene ice epoch, ice age cycles have been notably longer and colder with year-round polar ice and intervening interglacials also being longer. In perspective with our 3,000-mile journey across I-80, the Pleistocene ice epoch began just 2 miles plus 317 feet from the end of the journey, or very near the western end of the I-80 Bay Bridge approaching San Francisco. Recent ice age cycle lengths (~110,000 years) would equate to just under 500 feet of the 3000-mile journey.

There is considerable evidence across the planet that some ice ages began precipitously[41] and did not develop slowly with gradually cooling global temperatures. Indeed, strong evidence supports the view that ice ages can begin so quickly that life forms have no opportunity to adjust to rapid and dramatic cooling. Over some parts of the planet, including areas that were experiencing semitropical conditions (the current climate of South Florida) at the onset of an ice age, mammals have been found to have frozen in ice so rapidly that food in their mouth and digestive tract had no time to putrefy. Such dramatic change is clearly unrelated to atmospheric CO_2.

Ice ages are separated by relatively brief moderate climate periods known as *interglacials*.[33] During the coldest ice age cycles, continental glaciers retreat but do not vanish (today Greenland and Antarctica are covered by continental glaciers). While still much colder than Earth's typical climate, interglacials are relatively short moderate climate breaks separating much longer bitter cold ice age cycles within ice epochs of ice eras.

Earth's climate is currently in the *Holocene* interglacial of an ice age cycle within the *Pleistocene* ice epoch of the current *ice era*.

Relatively mild, global average temperatures during interglacials still average 7.5°C (13.5°F) *colder than* Earth's typically much warmer climate, making interglacials *much colder* than Earth's typical climate.

Polar land masses (e.g., Greenland and Antarctica) often retain nearly all of their continental glacial cover during interglacials, a condition that does not exist during Earth's typically much warmer climate when *no year-round ice exists at sea level anywhere on the planet*.[38]

During the shorter ice age cycles of the early Pleistocene ice epoch, year-round polar ice was rarer, and when it did exist, it was far less extensive than it has been for the past one million years.

Because humans have never experienced Earth's typically much warmer climate, they have no real perspective with which to assert what global temperatures should be. It is the height of self-deceptive hubris to claim that humanity's puny net addition of CO_2 to the atmosphere from fossil fuel combustion is a force comparable to routine natural climate change forces.

Interglacial climate is neither characteristic of the bitter cold climate of ice ages nor of Earth's typically much warmer climate when not enduring an ice era. In short, interglacial climate is characteristic of at most one-half of 1% of Earth's recent 3.5-billion-year climate history, making interglacials the *least characteristic* of Earth's major climate regimes and a highly inappropriate basis for claiming what Earth's climate *should* be.

Yet based on a slight natural climate warming that is well within typical variability of Earth's natural climate perturbations and completely normal for an interglacial, the prosecution claims such warming is evidence that human activity is laying the foundation for "unprecedented" and potentially "catastrophic" or even "existential" climate warming! Really? On what basis is that claim asserted? What is the prosecution's standard for normal or typical climate? What degree of climate variability does the prosecution consider abnormal, and why?

Modern humans arose entirely during half of the Holocene interglacial of an ice age cycle in an ice epoch of an ice era.

In short, current climate is *anything but* typical for Earth and is certainly not a sound basis for asserting what climate should be. The jury will discover there is no evidence in any record that atmospheric CO_2 concentration has ever had any direct causal relationship with either climate change or the duration and intensity of ice ages and interglacials.

The current Holocene interglacial began 10,700 years ago. In perspective with our 3,000-mile coast-to-coast journey, the Holocene Interglacial began just 48'–5" from the end of the journey (present time), at most just *three car lengths* from the end of I-80 in San Francisco.

No evidence exists to support any claim that CO_2 growth or decline *has ever caused an ice age or interglacial either to begin or to end.*

Warm and Cold Periods

Lasting several hundred years, a climate optimum is also called a warm period (ex., *Minoan*, *Roman*, and *Medieval*), whereas a distinctly colder climate spanning hundreds of years is a cold period, the most notable being the recent and aptly named Little Ice Age[11] (LIA). Note the use of optimum (most favorable, most advantageous, etc.) denotes a *warm* climate period. Which climate would the jury prefer? Another ice age or Earth's typical very warm climate?

The following description is from *Atmospheres*[42] in the *Planet Earth* book series:

> Europe experienced the Little Climate Optimum[††] from about 800 to 1250 A.D.
>
> This was the age of the Vikings, when Norsemen not only invaded northern Europe but expanded their territorial domain to encompass Iceland and Greenland; Leif Ericsson and others are thought to have pressed even farther west to America. Greenland was named when its shores were indeed verdant, and the Norse settlers were able to raise oats, barley and rye. The colony grew until it comprised some 3,000 settlers living on 280 farms. The Little Climate Optimum was so warm that vineyards flourished in England, producing wines that supposedly rivaled those of France.
>
> Around 1250, the balmy days of the Little Climate Optimum began slowly but surely to wane. England once again became inhospitable to wine grapes, and on the Continent, vineyards that had flourished on hilltops had to be moved to lower, more protected sites. Long stretches of wet weather introduced a protracted cool period, and the particularly sodden decade of 1310 to 1320

[††.] Commonly known as the *Medieval Warm Period* (MWP).

brought terrible suffering to England and northern Europe…

Meanwhile, ice floes began to clutter the waters around Iceland, hindering access to Greenland. Soon, ships were unable to reach the Greenland colonists, farming became impossible and the settlement withered. In 1492 the Pope expressed concern that no bishop had been able to reach his Greenland flock for 80 years; he did not know that the last of the settlers had died by 1450…

The cold that turned Greenland into a frigid wasteland afflicted parts of Europe in the 15th century. Temperatures began a decline that was not dramatic—they were, on the average, only 1° to 2°F below those of 1200—but winters became longer and more severe, and summers were cooler and shorter. This climate phase would last more than 300 years and would come to be known as the Little Ice Age.

Asia was not spared the Little Ice Age. The cold settled over Japan and China in the 10th century and lasted until the 14th century. Exactly why the large-scale climate shift traveled slowly around the globe in this manner rather than descending over the entire Northern Hemisphere at once remains a puzzle to climatologists.

Very gradually, imperceptibly at first, the cold began to slacken; during the 19th century, average temperatures increased a degree or two in the northern temperate latitudes. In the French Alpine village of Argentière, a glacier that had been pushing into the community's streets stopped in the 1850s, and soon began to retreat. Growing seasons lengthened and the Arctic pack ice retreated. The trend toward more benign weather lasted until World War II…

Around the middle of the 20th century, temperatures in the Northern Hemisphere once more began to inch downward. Between 1940 and 1965 the Northern Hemisphere cooled by about 0.5°F, on the average, and from 1951 to 1972 water temperature in the North Atlantic declined steadily…

Think about the above description of the Medieval Warm Period (Little Climate Optimum). Is current global climate remotely as warm as the MWP climate that melted Greenland's glacial ice to the extent the name "Greenland" was appropriate and 280 farms had been established by the 15th century?

Ironically, the referenced 1940 to 1965 cooling accompanied a rapid increase in atmospheric CO_2! In perspective with the I-80 journey across the US, the Little Ice Age ended roughly 170 years ago, putting the end of the LIA at less than 9-1/4" from the end of the 3,000-mile journey! Yes, just 9-1/4". A shoe length!

In perspective with 3.5 billion years of past climate, the past 50 years of climate change is equivalent to less than the length of a credit card along the 3,000-mile journey.

The Solar Connection

Scientists really don't know what triggers sudden dramatic changes from Earth's typically very warm climate to very cold ice eras. However, most scientists acknowledge a strong solar connection with climate change, particularly during ice age-interglacial cycles of ice epochs.

Climate and climate change are impacted by two distinctly different sources of variable solar irradiance at Earth's surface:

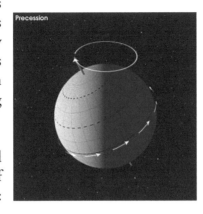

1. Earth's orbital changes
2. Solar variability

Earth's Orbital Changes

Orbital changes affecting solar irradiance at Earth's surface are (1) axial precession, (2) axial obliquity, (3) orbital eccentricity, and (4) orbital obliquity.[43]

Axial precession is the slow coning rotation of the direction of Earth's axial tilt around the plane of day-night separation (see illustration above, from *NASA Earth Observatory*). This movement is represented by the white circular motion path traced by the red arrow at the northern tip of Earth's axis of rotation. Note the pole of the sphere illustrating the movement (*precession*) of the axis of Earth's rotation.

Axial obliquity is the magnitude of the tilt of the Earth's axis off the plane of day-night separation. The angle of obliquity is the angle between the red arrow-tipped axis and the vertical (an imaginary vertical axis at the center of the white precession circle). When the angle becomes zero, there is no seasonal climate change beyond that attributed to orbital eccentricity.

Orbital eccentricity affects the shape of Earth's orbital rotation around the Sun. While often thought of as circular, Earth's orbit around the Sun is slightly elliptical (an elongated circle). Eccentricity is a measure of how much an ellipse is elongated (eccentricity = 0 for a perfect circle). Earth's orbital eccentricity is very small and varies between 0.0034 at minimum (nearly a perfect circle) to a maximum of 0.0580. Earth's current orbital eccentricity is 0.0167, a very slightly elliptical orbit (too slight to illustrate).[44] These changes slightly affect the distance from Earth to the Sun and will, to some degree, impact the strength of solar irradiance as Earth's orbital position changes.

Orbital obliquity is the tilt of the plane of Earth's orbit around the Sun, another variation that affects solar irradiance at Earth.

Cycling at different rates, these orbital characteristics will change over time. Such change affects the times when Earth is closest to the Sun during Northern Hemisphere summer (and most distant during winter), leading to greater seasonal variability of Earth's climate. Orbital characteristics will combine to create periods of warmer and colder climate.

> "The largest difference in [climate change] forcing between and during different interglacials lies in the latitudinal and seasonal pattern of incoming solar radiation…which is controlled by the three astronomical parameters precession, obliquity, and eccentricity."[40]

The citation above credits natural forces (Earth's orbital changes) with being the dominant factor affecting ice age-interglacial cycles.

Not greenhouse gases or carbon dioxide but the effect Earth's natural orbital variations have on solar heating of Earth's surfaces.

While these particular orbital changes may not be significant over periods as short as a few hundred years, other natural factors altering solar irradiance are responsible for nontrivial atmospheric temperature change. Variable solar activity can be a potent influence on Earth's multidecadal and multi-century climate variability.

Variable Solar Activity

The prosecution persistently maintains[45] varying solar irradiance *is not* a significant contributor to climate change:

> The evidence for human influence on the climate system has grown since the IPCC Fourth Assessment Report (AR4). It is extremely *likely* that more than half of the observed increase in global average surface temperature from 1951 to 2010 was caused by the anthropogenic increase in GHG concentrations and other anthropogenic forcings together.[45]

As testimony will clearly show the jury, there is not a scintilla of real-world evidence to support this prosecution claim.

In *Climate Change: The Facts*,[46] the following passage on page 62 describes how the prosecution *ignores* the obvious:

> IPCC authors have also failed to disclose or to explain that the measurement of total solar irradiance is *confounded by our current inability to determine its absolute value.* It is surely important

to know whether the mean value is 1360, 1361, or 1365 Wm^{-2} because without knowing how the mean climatic state is determined it would be impossible to confirm how the climate system is actually changing. The scientific importance of this indeterminacy is also clear if one considers that, according to the IPCC's 2013 *Fifth Assessment Report*, the *entire influence of humans on the climate since 1750 is a mere 2.3 Wm^{-2}.*

The 2013 paper by Soon and Legates, published before the IPCC's paper cut-off deadline, shows that *a reconstructed history of solar irradiance can explain the changes in the Equator-to-Arctic surface temperature gradient from 1850-2012.* Scientifically, this result is important for understanding climate dynamics because the Equator-to-Arctic temperature gradient has long been suspected as a key driver of the Earth's climate. *The IPCC, which purports to review all relevant scientific literature, makes no mention of this important result.*[47] [*emphasis* added]

According to these papers,[46–47] a firm value for the Sun's true mean irradiance, estimated to be between 1360 and 1365 Wm^{-2} (watts per square meter), has not been established to a precision within 5 Wm^{-2}. Note that this precision is not the instrument measurement precision, which can be much smaller.

Yet, the IPCC's 2013 Fifth Assessment Report, AR5, concludes the *entire human impact on climate* over the past 269 years (1750 to 2019) is thought to be a mere *2.3 Wm^{-2}* or *less than half the acknowledged uncertainty (5 Wm^{-2})* associated with measuring solar irradiance! *The human impact on climate is buried within the accuracy of knowing the* *actual value of solar irradiance!* Under the circumstances, trying to predict future climate is like trying to read by the light of a lamp with no lightbulb (not very bright).

The prosecution (IPCC) has always dismissed natural solar variability by claiming it has no meaningful impact on Earth's recent climate change. Yet if the error associated with knowing the real value of solar irradiance is more than twice the "entire influence of humans on the climate since 1750…"[46] then upon what possible evidence does the prosecution base its claim that human activity is governing climate change? None. It bases its claim entirely on its blind devotion to a flawed theory and its dubious climate simulation models based on that theory.

When published studies[46–48] conclude climate change could be entirely caused by solar variability (variable irradiance), the prosecution ignores them!

The period cited by the IPCC[37] (1951 to 2010) as having "more than half the observed increase in global average surface temperature" caused by human activity is fully examined by the jury in Chapter 4.

Figure 3-3 is based on Figure "SPM.3 on page 6" of the cited IPCC Fifth Assessment Report. This figure presents the prosecution's estimated "contributions to observed surface temperature change over the period 1951–2010." Recall that during 1951 to 2010, solar activity peaked (a solar grand maximum; a peak of peaks), a natural force the prosecution's chart (Figure 3-3) indicates as having no meaningful impact!

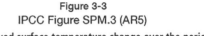

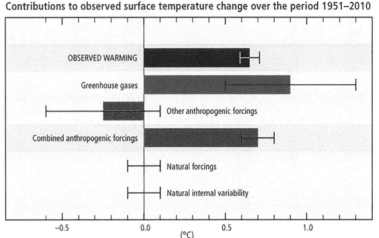

Figure 3-3
IPCC Figure SPM.3 (AR5)
Contributions to observed surface temperature change over the period 1951–2010

Source: IPCC Climate Change 2014 Synthesis Report, Summary for Policymakers, page 6, Figure SPM.3

An associated stretched "H" line illustrates the prosecution's range of uncertainty for each bar. Note the very large uncertainty associated with "Greenhouse gases" and "Other anthropogenic forcings." Greenhouse gases uncertainty is nearly as large as the claimed contribution from greenhouse gases! This IPCC chart appears to be little more than a convenient fantasy.

Because it is impossible for anyone to actually measure the contribution from greenhouse gases, the contribution can only be estimated by theory-based climate simulation models whose performance to date has been questionable.

Looking at Figure 3-3, the basis for the large "Greenhouse gases" bar is certainly dubious as it is directly tied to the prosecution's climate change *theory* that CO_2 and temperature records will reveal to be spectacularly flawed. It is entirely fitting that the greatest degree of uncertainty is associated with the prosecution's confidence in the forcing associated with greenhouse gases. Compelling defense testimony in Chapter 4 will show jurors why this chart's forcings for both "Greenhouse gases" and "Combined anthropogenic forcings" may be more akin to science fiction than scientific fact.

Figure 3-4 (next page) shows the strong relationship between climate and solar activity. Note the coincidence of late 20th-and early 21st-century warming with the modern solar grand maximum peaking during much of the 1951 to 2010 timeframe.[25, 45] Coincidence or causation? Jurors might reasonably question how it is that the strongest solar grand maximum in nearly 800 years could have the astonishingly trivial impact claimed in the prosecution's Figure 3-3 for "Natural forcings"!

From *Modern Solar Grand Maximum Ends: 'Little Ice Age' Cooling Coming!*:[25]

> During the 20th and early 21st centuries, Earth's inhabitants have enjoyed an epoch of very high solar activity that is rare or unique in the context of the last several thousand years. The higher solar activity and warmer temperatures have allowed the planet to briefly emerge from the depths of the successive solar minima periods and "Little Ice Age" cooling that lasted from the 1300s to the early 1900s. Unfortunately, solar scientists have increasingly been forecasting a return to a solar minimum period in the coming decades, as well as the concomitant cooler temperatures… In several newly published (2017) papers, scientists have suggested that a substantial deterioration into solar minimum conditions and global cooling may be imminent.

Also, in *Do Models Underestimate the Solar Contribution to Recent Climate Change?*[49]

> Current attribution analyses that seek to determine the relative contributions of different forcing agents to observed near-surface temperature changes underestimate the importance of weak signals, such as that due to changes in solar irradiance. Here a new attribution method is applied that does not have a systematic bias against weak signals.

> It is found that current climate models underestimate the observed climate response to solar forcing over the 20th century as a whole, indicating that the climate system has a greater sensitivity to solar forcing than do models.

To summarize these profound observations:[25, 49]

1. Solar variability has had a *significant impact* on climate over the past thousand years.
2. Biased theory-based climate models deliberately downplay solar variability's role in climate change while vastly overestimating the greenhouse gas contribution to climate change.

Should jurors be surprised to learn that climate models are unreliable? Has the prosecution attempted to hide evidence exculpatory to atmospheric carbon dioxide while simultaneously failing to disclose to the jury a realistic alternative basis for recent climate change?

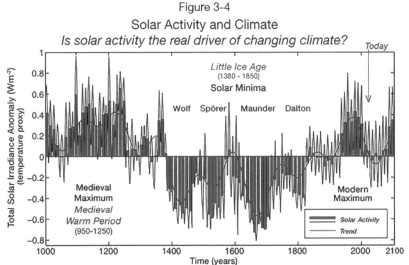

Figure 3-4
Solar Activity and Climate
Is solar activity the real driver of changing climate?

Based on: Herrera, et al, (July 11, 2014), *Reconstruction and prediction of the total solar irradiance: From the Medieval Warm Period to the 21st centruy*, Figure 9a, New Astronomy.

Figure 3-4 clearly illustrates these cycles have historically shown a far more consistent relationship and better correlation with Earth's climate change (represented by the Medieval Warm Period, the Little Ice Age, and recent modern warming) than does the prosecution's theorized relationship between CO_2 change and climate change. This will become even more evident in Chapter 4 when the jury examines the veracity of the prosecution's allegation that growing atmospheric CO_2 is a strong climate change force causing significant global temperature warming and long-term global climate warming.

These examples illustrate, both individually and combined, how orbital variability and solar variability are capable of producing far greater climate variability than any effect claimed for atmospheric CO_2 growth.

It is unlikely that orbital and solar variability are of sufficient power to spawn or terminate *ice eras*. For one thing, the onset time and duration of ice eras are vastly longer than cycles of orbital and solar variability. For another, the magnitude of climate cooling needed to trigger an ice era is well beyond the range of variations any known combination of orbital and solar variability is likely to produce.

However, Figure 3-4 makes undeniably clear that cycles of orbital and solar variability have had a direct and significantly well-correlated impact on the timing of warm and cold periods during the current interglacial.

While our Sun is a relatively stable star, it has instabilities manifested by various cycles. Evidence of these cycles of solar variability is the coincidence of observed solar variability with climate change as depicted in Figure 3-4. Solar minima are typically indicated by solar "quiet" episodes associated with reduced sunspot activity. Note particularly that there is *no evidence that greenhouse gases have any impact* on the climate change shown in Figure 3-4.

The jury should bear in mind that climate is affected by a number of disparate, unrelated forces, many of which have an influence that can be modified (either enhanced or suppressed) by other forces. Suppose, for example, a solar grand maximum occurs during a period when orbital conditions align so that Earth's orbit carries it closest to the sun during summer. Such an alignment of two warming forces would combine to produce a warmer summer climate than either force could achieve on its own. Similarly, suppose a solar grand minimum were to occur when orbital conditions align so that Earth's orbit carries it farthest from the sun in winter. Such an alignment of two cooling forces would combine to produce a colder winter climate than either force could achieve on its own.

These examples illustrate how orbital variability combined with solar variability can produce far greater climate variability than any effect observed from atmospheric CO_2 growth.

The recent Little Ice Age coincided with a solar grand minimum, a period of very low solar activity whose lowest point occurred in the late 16th century but that began in the 12th century and lasted until the mid-19th century. Warm periods, on the other hand, are characterized by frequent sunspot activity that is associated with warmer temperatures. In terms of a human lifespan, these changes are relatively slow with unpredictable year-to-year and decade-to-decade variability. Reverses in trends occur with some frequency. Note the variability of solar activity between the years 1400 and 1850 of the LIA.

Looking at Figure 3-4, would scientists living near the end of the Medieval Warm Period anticipate future warming or cooling? On what basis? What about those scientists living near the end of the Little Ice Age? The evidence clearly shows that future temperature trends *are not* foretold by current climate. Had scientists known that sunspot count can be a fair indicator of climate change, they might have noted any significant change in the number of sunspots. Yet even then, it is difficult to predict future sunspots based solely on a history of past sunspots.

Science has yet to find an explanation for the sudden onset of very deep climate cooling episodes ranging from ice eras to ice age cycles, whose multiple appearances are indubitable. Understanding this and the geologic evidence of natural climate and climate change over the past 3.5 billion years should help jurors put climate change concerns into proper perspective.

Recall the geologic evidence that Earth's most typical climate is about 14°F warmer than any climate modern humans have experienced. With that in mind, and given the relatively trivial warming over the most recent fifty years of post-LIA climate during the Holocene interglacial of the Pleistocene ice epoch of the current ice era, does it seem the least bit reasonable for the prosecution to claim recent climate change is a rational basis upon which to warn that relatively slight future climate warming will be "unprecedented"?

In perspective with historic natural climate change, is there any reasonable basis upon which the prosecution can legitimately claim growing CO_2 "is extremely *likely*"[45] to be the cause of recent observed climate warming?

Summary: Climate Perspective

1. Earth's typical climate during 90% of the past 3.5 billion years is characterized by the climate that prevailed when dinosaurs lived and is much warmer than any climate humans have ever experienced.
2. An ice era is a period of extreme cold climate deviating from Earth's typical climate and averaging fifty million years duration.
3. Earth's climate has experienced only seven ice eras that collectively span 350 million years (or just 10%) of the past 3.5 billion years since life first appeared on Earth. During ice eras, global average temperatures drop by as much as 10°C (18°F) below Earth's typically much warmer climate.
4. Year-round polar ice caps *only exist* during the coldest climate regimes of ice epochs embedded within ice eras.
5. Ice epochs, roughly spanning several million years, are the coldest climate regimes within an ice era.
6. Ice ages are the coldest climate regimes within an ice epoch of an ice era.
7. Interglacials spanning thousands of years are relatively mild climate interludes separating bitter cold ice age cycles that typically range from 40,000 to 100,000 years duration. Interglacials, while milder than ice ages, are *much colder* than Earth's typical climate when Earth is not in an ice epoch of an ice era.

8. Earth's climate is currently in the Holocene interglacial of an ice age cycle within the Pleistocene ice epoch of an ice era.

9. Current global climate is averaging about 13.5°F colder than Earth's typical climate.

10. Climate during an interglacial of an ice age cycle within an ice epoch of an ice era is *the least typical major climate regime* on Earth. Interglacials are much colder than Earth's typical climate and more moderate than the longer-duration coldest climate of ice ages.

11. There is a reasonable scientific basis for acknowledging solar and orbital variability are capable of combining to play a significant role in multi-century warm and cold periods, as well as interglacial and ice age cycles.

12. Solar variability plays an important role in multi-decade climate change and is known to be real and significant.

13. Scientific certainty governing natural causes of warm and cold periods, interglacials, ice age cycles, ice epochs, and ice eras is not well-established, yet such climate regimes are natural for Earth's climate.

14. Without scientific certainty, meaningful projections are impossible and run the risk of being misleading with disastrous consequences.

The 550 Million Years of Climate and Atmospheric CO_2 Change

A clear understanding of the historic relationship between Earth's climate and its atmospheric CO_2 is an essential precondition for jurors to reasonably weigh the evidence to be presented in Chapter 4.

While human knowledge may change, science does not. Past relationships between atmospheric CO_2 and climate are valid indicators of future relationships. Figure 3-5[39] shows Earth's climate and atmospheric CO_2 history based on geologic evidence over the past 550 million years. This evidence should prove immensely helpful to jurors seeking to better understand the nature of any relationship between Earth's atmospheric CO_2 concentration and its climate and, in particular, the relationship between *changing* atmospheric CO_2 and climate *change*. The insights gained should help jurors better assess the validity of prosecution's theorized claims.

As historic geologic evidence clearly testifies, Earth's atmospheric carbon dioxide *cannot possibly* be the source of dramatic changes that both initiate and terminate ice eras. Consequently, *it is implausible* to assume CO_2 could trigger or end ice epochs, ice age and interglacial cycles, or any other climate cycles spanning mere centuries or even decades. Note that on the scale of 550 million years, temperature is actually a measure of climate. Every discernible temperature point on the temperature curve represents about one million years, and it would be hard to argue that a temperature representing one million years does not reflect climate.

Despite this evidence, the prosecution *asserts* its greenhouse gas climate change theory that—contrary to the evidence—claims (1) atmospheric CO_2 growth "traps" more heat, impeding Earth's cooling and raising Earth's surface temperatures and (2) atmospheric CO_2 decline will "trap" less heat, allowing Earth to cool more readily, lowering Earth's surface temperatures. In short, the prosecution claims growing atmospheric CO_2 causes climate warming and declining CO_2 causes climate cooling. Yet the evidence revealed by Figure 3-5 clearly contradicts that theory.

If the prosecution's theory is valid and changing atmospheric CO_2 is a *strong force* causing climate change, then the evidence of that change (warming or cooling global climate) should be virtually coincident with changing atmospheric CO_2.

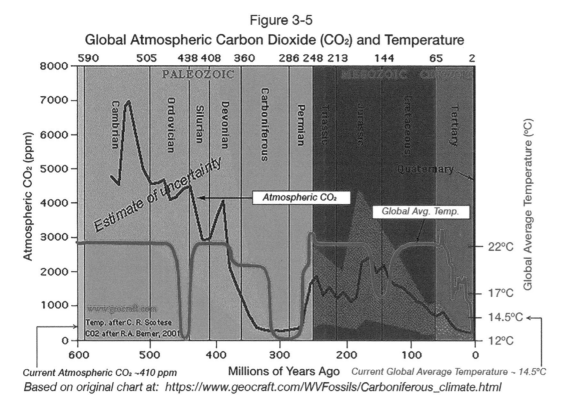

Figure 3-5
Global Atmospheric Carbon Dioxide (CO₂) and Temperature

Current Atmospheric CO₂ ~410 ppm
Based on original chart at: https://www.geocraft.com/WVFossils/Carboniferous_climate.html

Three ice eras are depicted. The current ice era began 60 million years ago; an earlier ice era bottomed about 290 million years ago and lasted 50–60 million years; and a relatively short bitter cold ice era spanning a brief 25-30 million years was at its coldest about 450 million years ago. A period of lesser cooling of just -5°C (-9°F) lasting 40 million years appeared about 160 million years ago but was not sufficiently cold to spawn ice age cycles. This lesser cooling occurred during the dinosaur era and was sufficiently warm to not be catastrophic for dinosaurs.

Worth noting is that throughout the 550-million-year record while climate was varying between the typically warm and relatively stable to relatively brief bitter cold ice eras, atmospheric carbon dioxide was all over the place. Most notably, during the earliest of the identified ice eras, *atmospheric CO₂ grew ~500 ppm* as *global climate plunged* into the depths of a deep ice era. Then as *atmospheric CO₂ declined* by 1,500 ppm, *global climate warmed* to typical very warm levels.

Long-term dramatic changes in CO₂ should be manifestly evident in the long-term climate record. Moreover, absent a stronger mitigating force, there is no basis in the prosecution's theory for climate change to lag behind CO₂ change over any meaningful period of time.

If climate change theory were valid, the degree of climate change should be reasonably proportional to the degree of atmospheric CO₂ change (greater and lesser changes in global atmospheric CO₂ should correspondingly produce greater and lesser changes in global temperature).

Referring to Figure 3-5, note that for most of the 550 million years for which both CO₂ and temperature (a measure of climate) are shown, climate is predominantly Earth's typical very warm climate of the past 3.5 billion years, a climate warmer than anything humans have experienced.

Despite the large uncertainty in the CO₂ estimates as time recedes into the past, this evidence is remarkably contradictory to the dictates of the prosecution's climate change theory.

Between 140 mya and 70 mya, when atmospheric CO₂ (black line) fell consistently from ~2,200 ppm to ~800 ppm, climate sharply warmed by 5°C (9°F), then leveled off! Exactly how is that possible if, as the prosecution alleges, changing atmospheric CO₂ is a strong climate change force?

While the prosecution is entitled to its theory, *the scientific method* informs us the prosecution is not entitled to merely *assert the validity* of that theory, particularly in the face of compelling contradictory evidence.

Table 3-2
550 Million Year Geologic History
of
Atmospheric CO₂ and Climate

Millions of Years Ago (mya)	Atmospheric CO₂ (ppm)	Global Climate (degrees C)
550	4603	22.0
525	6635	22.0
500	4634	22.0
475	4246	21.9
450	4416	12.2
425	3500	22.0
400	3506	22.0
375	1713	20.3
350	779	20.0
325	345	19.8
300	286	12.3
275	368	13.0
250	1844	22.8
225	1509	22.0
200	1260	22.0
175	2363	21.9
150	2039	16.7
125	1673	20.4
100	1312	22.0
75	807	22.0
50	805	22.8
25	345	17.6
0	400	14.5
mean	2147	19.7
median	1673	21.9
correlation coefficient	(CO₂, T) = 0.29 ΔCO₂, ΔT) = 0.10	

Original source: geocraft.com
Climate: C. R. Scotese, CO₂: R. A. Berner (2001)

Table 3-2 is created by estimating atmospheric CO_2 and climate from Figure 3–5. Correlation coefficients between both CO_2 and climate and CO_2 *change* and climate *change* are calculated using the tabulated estimates. Over the past 550 million years, a very low correlation coefficient (0.29) exists between atmospheric CO_2 and global climate. There is even less correlation between *atmospheric CO_2 change* and global *climate change* (0.10 correlation coefficient). These very low correlations offer further compelling evidence the prosecution's climate change theory is invalid.

The prosecution's theory dictates that growing atmospheric CO_2 (the action) is responsible for climate warming (the reaction). In *Role of Greenhouse Gases in Climate Change*,[50] the authors note: "While a parallelism between two separate quantities does not prove that the two are causally related, *the lack of parallelism proves that they are not causally related*" [*emphasis* added]. A causal relationship creates correlation; lack of correlation *assures no causal relationship can exist.*

To help visualize the inconsistency of climate change theory with the evidence, the record in Table 3–2 is used to create Figure 3–6 (next page) that visually indicates when global climate change and changing global atmospheric CO_2 are consistent with theory (green bars *above* the horizontal line) and when they contradict (are inconsistent with) theory (red bars *below* the horizontal line).

If the 25-million-year change in either global atmospheric CO_2 or global climate is trivial while the corresponding change in the paired data is large, the comparison is tallied as *inconsistent* with greenhouse gas climate change theory.

More than 69% of the time, a contradiction to the prosecution's greenhouse gas climate change theory is indicated. This is consistent with the very low correlation coefficient (0.10) between changing atmospheric CO_2 and changing climate in the geologic record.

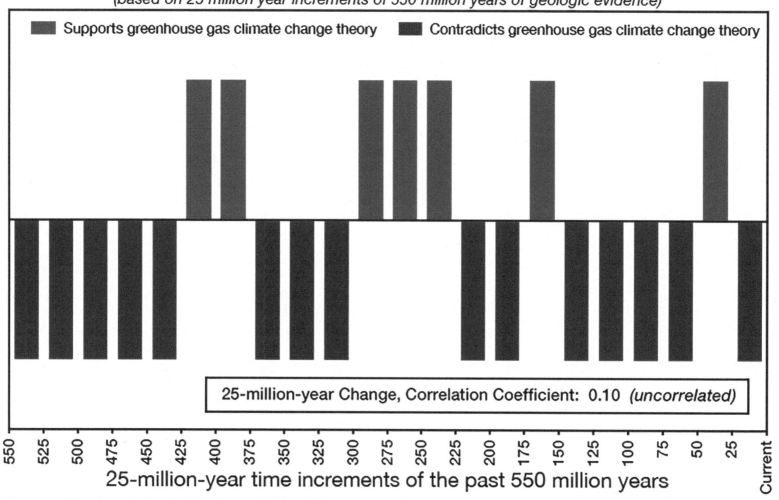

Figure 3-6

Does *Direction of* Atmospheric CO_2 Change Match *Direction of* Climate Change?

(based on 25 million year increments of 550 million years of geologic evidence)

Supports greenhouse gas climate change theory Contradicts greenhouse gas climate change theory

25-million-year Change, Correlation Coefficient: 0.10 *(uncorrelated)*

550 525 500 475 450 425 400 375 350 325 300 275 250 225 200 175 150 125 100 75 50 25 Current

25-million-year time increments of the past 550 million years

Source: *Climate and the Carboniferous Period, Temperature after C.R. Scotese, CO2 after R.A. Berner, 2001 (GEOCARB III), http://www.geocraft.com/WVFossils/Carboniferous_climate.html*

Chapter 4 will examine testimony of both contemporary and ice core records for any evidence of compatibility with climate change theory.

Summary: 550 Million Years of Climate and Atmospheric CO_2 Change:

1. Contradicting the prosecution's greenhouse gas theory, the 550-million-year geologic record of global atmospheric CO_2 and global climate shows neither a consistent relationship nor a plausible causal relationship exists. If the prosecution's theory had merit, then the geologic record would support it. But it doesn't. There simply is no evidence that the concentration of atmospheric CO_2 has any relationship with Earth's global climate. No relationship means no causation is possible.

2. The geologic record shows that during 30 million years between 465 and 435 mya, contrary to prosecution's theory, climate cooled 10°C then warmed 10°C while atmospheric CO_2 changed dramatically *the opposite*, growing 500 ppm while *climate dramatically cooled,* and then falling 1,500 ppm while climate *warmed*. Then between 150 mya to 110 mya, while CO_2 fell by 700 ppm, global climate warmed by 5°C. From 320 mya to 300 mya, climate cooled 8°C while CO_2 was *unchanged*. These examples are dramatically contradictory to the demands of the prosecution's theory.

3. If changing atmospheric CO_2 actually does cause a corresponding change to global climate, then the two records will be *highly correlated* and their correlation coefficient will be somewhere in the range of 0.8 to 1.0. Yet over the past 550 million years, the correlation coefficient between CO_2 and climate is a mere 0.29 and between CO_2 *change* and climate *change* is a trivial 0.10. No correlation means no causation can exist.

Perspective on Atmospheric CO_2 and Climate Change

Knowing the average and maximum atmospheric CO_2 concentration for a variety of climate periods should help jurors gain a proper perspective from which to examine the lack of any consistent relationship between atmospheric CO_2 and climate in Earth's historic climate record.

The climate periods chosen to illustrate how these measures have ranged are: (1) the warmest climate over the past 550 million years, (2) the ice era climate of the past 60 million years, and (3) the coldest climate during the 2.4 million years of the current Pleistocene ice epoch.

A bar chart (Figure 3-7) visually compares this evidence in terms of the average and maximum atmospheric CO_2 for each of these periods. Note that the ice epoch is part of the ice era which, in turn, is part of the past 550 million years. Average CO_2 is estimated to be 2,147 ppm over the entire 550 million years, 461 ppm during the current ice era (60 million years), and 235 ppm during the current ice epoch (2.4 million years).

Why is this important?

- A mere 105 ppm change of atmospheric CO_2 over the past 128 years is not the least bit unusual and is well within the natural range of atmospheric CO_2.
- Both the range and average atmospheric CO_2 of warmer climate regimes show higher levels of atmospheric CO_2, but atmospheric CO_2 *change* is clearly independent of climate *change*, as is testified to by Figures 3-5 and 3-6. This indicates atmospheric CO_2 *can be indirectly influenced by climate*, but atmospheric CO_2 *cannot change climate*.

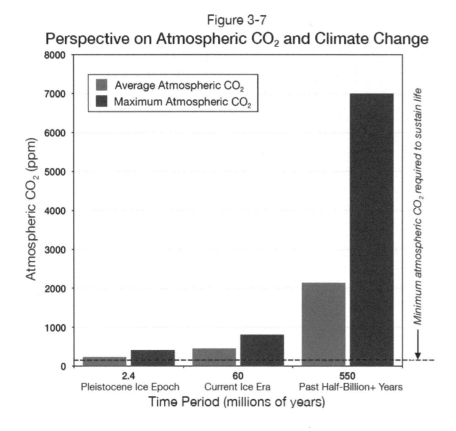

Figure 3-7
Perspective on Atmospheric CO₂ and Climate Change

For perspective, the dashed line near the bottom of Figure 3-7 indicates 135 ppm, the *minimum level of atmospheric CO₂ necessary to sustain life*.

Note that the 550-million-year record is characteristically *much warmer* than any climate humans have experienced, evidence that humans have never experienced what might be considered Earth's "normal" climate. It is arrogant to assume the climate typical of modern human existence, at the most, just 0.000001 (or 0.0001%) of Earth's existence, is a reasonable basis for believing what climate *should* be.

Don't be misled by the prosecution's false claims that recent CO₂ levels are dangerously high. As Figures 3-5 and 3-6 make abundantly clear, no causal relationship exists between atmospheric CO₂ *change* and climate *change* over the 550-million-year record.

While atmospheric CO₂ has been relatively low (at 810 ppm maximum) during the current ice era, between roughly 470 and 430 million years ago the coldest ice era of the past 550 million years occurred when atmospheric CO₂ was well above 4,000 ppm! This evidence flies in the face of the prosecution's greenhouse gas climate change theory.

In Chapters 4 and 5, the defense will further challenge the validity of the prosecution's climate change theory with additional compelling evidence found in contemporary measured records.

How Can We Know What Isn't Natural If We Don't Know What Is?

If asked to investigate the cause of climate warming over the past 100 years, an objective scientist would first ask:

- Is there any evidence climate change during the past 100 years is inconsistent with normal climate variability?
- What is known about natural climate change, the possible causes of such change, and the sensitivity of climate to each of those possible causes?
- For each identified natural cause, what evidence do we look for, and how is that evidence monitored?
- What portion of observed climate change since the late 19th century can reasonably be attributed to each of the identified natural causes?

If the answers to these questions were known, an objective scientist would then have a credible basis for identifying evidence for the existence of any significant climate change for which no natural cause is known. Under no circumstance is it legitimate to simply *assert* human activity is causing climate change and then manufacture supporting evidence using climate simulation models built on the basis of the very theory in question! Given scientists' limited state of knowledge about climate change and its causes, there remains the distinct possibility of as yet unidentified or improperly quantified natural causes.

If the four questions above do not have clear, unambiguous answers, then, in the absence of supporting evidence in the atmospheric CO_2 and temperature records, it is foolish to assert *any portion* of observed climate change is anything other than natural.

Why is that? Because climate change over the past 3–4 billion years is routine, dramatic, and typically *far more pronounced* than any climate change detected since the mid-19th century. Any claim that recent climate change is either unprecedented or unusual misleads the jury because not only is recent climate change precedented, it is also *minimal when viewed in perspective* with historic climate change!

By definition, nonnatural climate change implies some human factor. Before just assuming a nonnatural causation for any recent observed climate change, the vast history of naturally occurring climate change[51–52] needs to be examined to develop a familiarity with past climate change and a better understanding of its possible causes.

The Medieval Warm Period (MWP) and Little Ice Age (LIA) are the two most recent notable global climate perturbations. Figure 3-4 (page 43) shows strong evidence that each of those climate excursions was a product of solar variability. Yet prosecution agents like to claim both were not global on the specious grounds that little recorded evidence of worldwide scope exists. If these climate variations were triggered by solar events (and the evidence strongly indicates they were), *how could they not be* global in nature? Furthermore, just because the preponderance of evidence is found in developed areas of the Northern Hemisphere where most of the more advanced civilizations flourished doesn't mean the events weren't global in nature; it simply means the likelihood of having a clear record is very small in undeveloped areas, just as more tropical storms were discovered after satellites began looking for them. Consider too that the Northern Hemisphere has *twice the land surface* of the Southern Hemisphere, whose largest land mass is glacier-covered! One would expect there to be more climate evidence in the more heavily populated areas of the Northern Hemisphere.

The obvious relationship between solar activity and climate (Figure 3-4) belies any claim that these warm and cold events were merely local and not global. Both are highly correlated with solar activity and spanned hundreds of years, affecting widespread areas across the globe. While both the MWP and LIA are relatively minor climate change episodes when compared with ice age and interglacial cycles, both were longer, more pronounced, and far more significant than any observed modest climate variations since the mid-19th century following the end of the LIA. Wouldn't it behoove the prosecution to develop a better understanding of what both precipitated and terminated those well-established global climate change episodes rather than simply trying to dismiss them by misrepresenting their global scope?

A clear understanding of the scope and causes of natural climate change, together with historic records confirming such relationships, should be *a minimum prerequisite* to any legitimate claim that human activity is causing potentially catastrophic climate change.

Table 3-3
Contradictions to Greenhouse Gas Climate Change Theory
Based on 550 million years of geologic evidence

Coldest			Warmest		
mya	Temp (°C)	CO_2 (ppm)	mya	Temp (°C)	CO_2 (ppm)
450	12.2	4416	525	22.0	6635
300	12.3	286	425	22.0	3500
150	16.7	2039	250	22.8	1844
current	14.5	400	60	22.8	805

Based on: https://www.geocraft.com/WVFossils/Carboniferous_climate.html

The prosecution *began with a theory* blaming fossil fuels as the principal cause of all recent climate change. In a convoluted circular approach to scientific investigation, it then confined its activities to creating supportive studies and reports based on its theory and theory-driven climate simulation models in an attempt to justify the very theory upon which its evidence is based! That isn't science; it's self-deception on a grand scale.

The simple truth is that scientists cannot legitimately identify any portion of recent climate change as being unnatural because they cannot say what portion is natural. Changing climate doesn't have to be unnatural.

The prosecution alleges fossil fuel emissions of carbon dioxide are responsible for climate warming. Without first clearly establishing in the record the evidence that changing atmospheric CO_2 actually causes global climate change, that claim is both reckless and unsupported by evidence.

If the record cannot unequivocally establish atmospheric CO_2 change is a persistent force causing climate change, *then any source of atmospheric CO_2 change is irrelevant*. For this reason, the defense will avoid speculation in favor of examining the actual recorded evidence to determine whether atmospheric CO_2 change has a persistent causal relationship with climate change that would support the prosecution's theory.

If the record clearly confirms atmospheric CO_2 change *is* a detectable force for climate change, then before the prosecution's theory is considered valid, prudence would dictate that all possible natural causes of observed climate change be either eliminated or quantified. Then, and only then, does science provide an opportunity for the prosecution to consider the extent to which fossil fuel emissions accumulate in the atmosphere sufficiently to have a significant impact on climate change.

The prosecution's indictment of atmospheric CO_2 claims that modest global climate warming of the late 20th century is caused by the growth of atmospheric CO_2, claimed to be unusual and human-induced.

Does the evidence from the 550-million-year geologic record support or contradict the prosecution's theory?

Consider Table 3-3 (based on Table 3-2) showing four instances during the past 550 million years when global climate cooled by more than 5°C below 22°C and four other instances when Earth's warmest climate (22.0°C to 22.8°C) prevailed (*mya = million years ago*).

Note that, in defiance of the prosecution's greenhouse gas climate change theory, the two coldest climate regimes, each lasting millions of years, had both very high and very low atmospheric CO_2 (4,416 and 286 ppm, respectively). Similarly, the two warmest climate regimes, each also lasting millions of years, had both high and moderately low atmospheric CO_2 (1,844 and 805 ppm, respectively).

Ironically the very coldest climate regime (at 12.2°C, 450 mya) occurred at a time when atmospheric CO_2 was 2.4 to 5.5 *times higher* than during the very warmest climate regimes (60 mya and 250 mya), all spanning millions of years.

The very coldest climate regime at 54°F global average temperature was 4.1°F colder than at present, whereas the very warmest climate regime at 73.0°F global average temperature was 14.9°F warmer than at present. Yet the *coldest* climate occurred when atmospheric CO_2 was 5.5 *times higher* than it was during the very warmest climate regime (4,416 ppm versus 805 ppm).

Table 3-3's inconvenient truths are not exactly a ringing endorsement of the prosecution's climate change theory. In fact, the geologic record makes a mockery of the prosecution's climate change theory, rendering atmospheric CO_2 entirely irrelevant to climate change.

Some might quibble about the magnitude of some of these geologic estimates, but they should bear in mind the precise magnitude isn't as important as is the evidence relating CO_2 to climate and CO_2 change to climate change.

While geologic temperature estimates are very accurate, the error bound (shown in gray on Figure 3-5) for atmospheric CO_2 becomes larger as the estimate moves back in time. Nevertheless, even with the lower confidence in the estimate, the strong contradiction to the prosecution's climate change theory is still very apparent. The precision of the numbers might change, but the relationship between them will not. That relationship is the dagger in the heart of the prosecution's theory.

Of key importance to the jury is the evidence that over the most recent half-billion years, for periods lasting tens of millions of years, the most severely cold climate has occurred when atmospheric CO_2 is substantially higher than it is during climate periods that are much warmer.

Over the 550 million year geologic record, *there is no consistent relationship between* either (1) *atmospheric CO_2 and global climate* or (2) *atmospheric CO_2 change and global climate change.*

Global average CO_2 and global average temperature are uncorrelated (correlation coefficient, r = 0.29), while global CO_2 *change* and global temperature *change* are *very strongly uncorrelated* (r = 0.10) over the 550-million-year record. *Lack of correlation means no causation is possible.*

The prosecution's entire case is based upon a theory that atmospheric CO_2 change causes global climate change. Yet the geologic record offers compelling evidence contradicting this theorized relationship. According to the geologic evidence, the prosecution's theory appears to be little more than speculative fantasy deceptively festooned with the aura of authoritative science.

In the next chapter, *"The Scientific Method"* will guide the jury's examination of the evidence characterizing the observed real-world relationship between changing atmospheric CO_2 and global climate change found in the mid-19th to early 21st century contemporary records.

Summary: *Instructing the Jury—Background Information*

1. Carbon dioxide. A colorless, odorless, invisible gas essential for all life on Earth. Using photosynthesis, plants absorb CO_2 from the atmosphere and use the carbon to grow while giving off oxygen as a byproduct. CO_2 reacts to a *very limited range* of infrared (IR) wavelengths.
2. Climate perspective. Humans have never experienced Earth's typical climate when Earth is not in an interglacial of an ice age-interglacial cycle within an ice epoch of an ice era. Spanning just thousands of years, interglacial climate is

the rarest significant climate regime on Earth and is the least typical climate of Earth's major climate regimes.

3. Climate change and atmospheric CO_2. Geologic evidence strongly establishes no correlation between global average atmospheric CO_2 change and climate change. The observed lack of correlation over hundreds of millions of years strongly refutes any theory that claims a strong causal relationship exists between changing atmospheric CO_2 and global climate change.

4. Virtually identical, yet warmer, natural climate warming episodes (e.g., Minoan, Roman, and Medieval warm periods) during the past five thousand years are unrelated to atmospheric CO_2. Consequently, the prosecution is speculating when it alleges current climate warming is unnatural and is caused by changing atmospheric CO_2.

5. During the past 550-million-year geologic record, global atmospheric CO_2 and climate are uncorrelated (r = 0.29). During that same time period, atmospheric CO_2 *change* and climate *change* are strongly uncorrelated (r = 0.10).

6. In the 550-million-year geologic record, *there is neither a consistent relationship between atmospheric CO_2 and global climate nor a consistent relationship between atmospheric CO_2 change and global climate change*, rendering the prosecution's case against atmospheric CO_2 baseless.

7. Methane (CH_4), also known as natural gas, while having a strong reaction to IR radiating from Earth's surface, is only reactive over a tiny fraction of the IR spectrum. That same frequency range of outbound IR is fully absorbed by water vapor in the atmosphere, leaving methane, an extremely rare atmospheric gas, incapable of any meaningful contribution to atmospheric temperature.

CHAPTER 4

Testimony of Global Average Surface Temperature and Atmospheric CO$_2$

The prosecution alleges:

> Recent climate warming is principally caused by atmospheric CO$_2$ growth.

The testimony presented in this chapter is found in archives maintained by the US government for records of global atmospheric CO$_2$ and global average surface temperature (GAST) since the late 19th century.

Any conclusions reached on the basis of these records are not the author's, they are the self-evident conclusions from *looking out the window* at the testimony of nature's evidence as recorded by government scientists.

Should this testimony be consistent with the geologic evidence and *not support* the prosecution's belief that atmospheric CO$_2$ is a strong climate change force, then the prosecution's allegation that fossil fuel emissions are responsible for the 20th century growth of atmospheric CO$_2$ is entirely irrelevant.

This trial thus reduces to a single question:

Is Atmospheric CO$_2$ Growth and the Greenhouse Effect Causing Global Climate Change?

The prosecution's allegation:

> Anthropogenic "greenhouse gas emissions…of carbon dioxide…are extremely likely to have been the dominant cause of the observed warming since the mid-20th century." (IPCC *Climate Change 2014 Synthesis Report, Summary for Policymakers*)[2]

The two key components of this allegation:

- The dominant cause of atmospheric CO_2 growth is anthropogenic greenhouse gas emissions (principally from fossil fuel use).
- The prosecution's climate change theory alleges observed atmospheric CO_2 change (growth) is causing Earth's global climate to change (observed warming).

The key question the jury will examine is whether or not the observed evidence in nature supports this allegation.

To answer that question, the jury will use the scientific method to guide its examination of the prosecution's theory.

The 800,000-year Antarctic Ice Core Record

Before examining records maintained by US government agencies for atmospheric CO_2 and GAST, the jury will examine the Antarctic ice core record of the past 800,000 years of Antarctic climate.

Figure 4-1 (next page) displays the 800,000-year Antarctic ice core record for atmospheric CO_2 (in red) and temperature (in black). This graphic may appear familiar as portions of it covering the most recent 450,000 years have been widely circulated for years. This is the same portion projected onto a stage in Al Gore's book and film, *An Inconvenient Truth*. Gore attempted to use *the appearance of correlation* as evidence atmospheric CO_2 was causing the observed climate change. Gore misled both his readers and his film viewers. While not readily apparent in this graphic, Gore neglected to mention that higher resolution ice core analyses had revealed climate was typically changing on average *800 years before* CO_2 changed.

On the scale of ice core records, temperature equates to climate since thirty years of temperature change is considered a valid measure of climate and thirty years is far too small to detect on this graph's scale. Indeed, even 800 years (1/1,000 of the time scale) is barely detectable on this graph's scale. So it was easy for Gore to mislead his audience.

The prosecution objects to this record and claims that at some point atmospheric CO_2 takes over and becomes the force causing climate change.

While it is true that the 800-year lag is not universal (it *is* an *average*), nevertheless, it dominates the record. For the few exceptions noted by the prosecution (e.g., about 370,000 years ago at the depth of the ice age), the ice core record appears to show CO_2 rising before temperature rebounds. However, this is not a characteristic observation across the ice core record and remains unexplained. Furthermore, an isolated exception to typical observations does not constitute supportive evidence for any theory.

From *What does the Vostok ice core tell us?*[53]

> A key point is that CO_2 does indeed respond to temperature changes (through ocean outgassing/ uptake and changes in vegetation) and—in the context of the glacial cycles—is more properly a feedback, than a forcing. This, however, does not mean that it is not a greenhouse gas and that changes in its concentration in the atmosphere will not influence global temperatures…[53]

Figure 4-1

Ice Core Proxy Climate Reconstruction Temperature Proxy vs. Atmospheric CO$_2$
(800,000 years of Antarctic Ice Core Analysis)

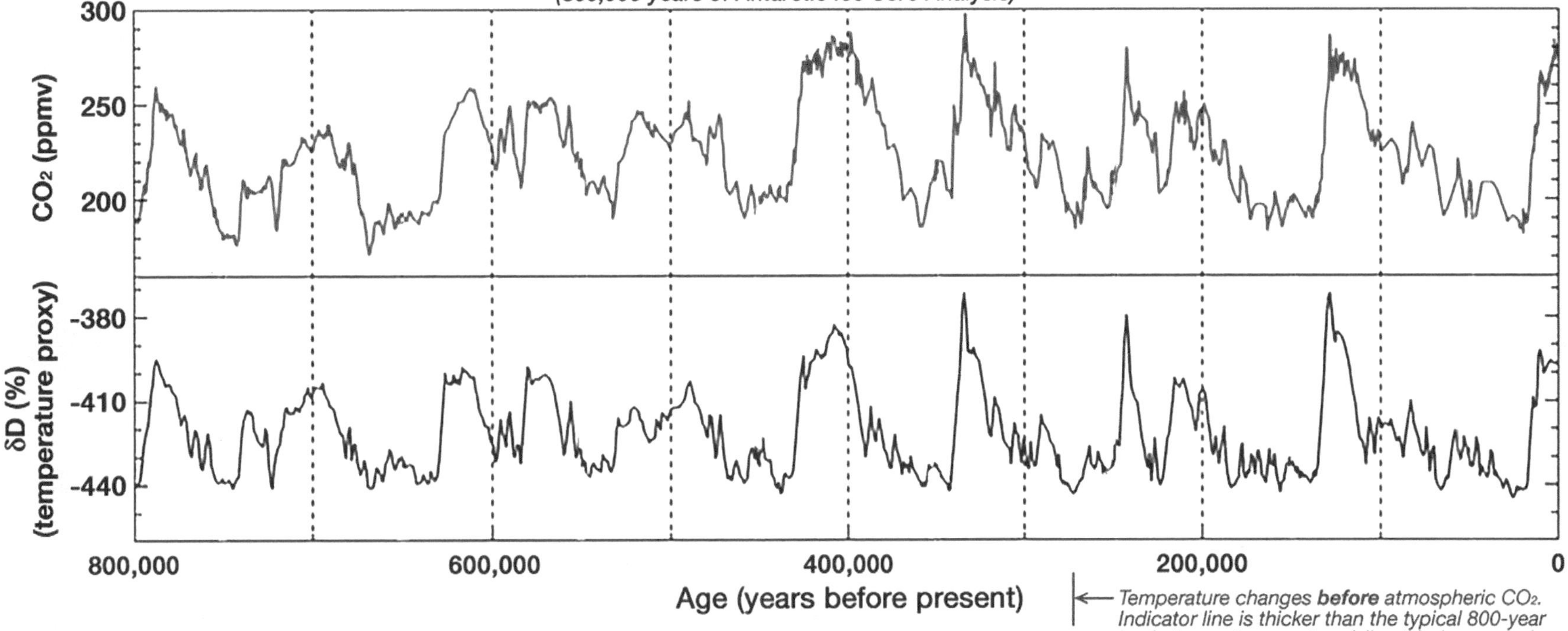

Temperature changes **before** atmospheric CO$_2$.
Indicator line is thicker than the typical 800-year lag between temperature (climate) change and subsequent atmospheric CO$_2$ change!

Source: Based on Figure 3 from "Ice cores and climate change" British Antarctic Survey (NERC) at (https://www.bas.ac.uk/data/our-data/publication/ice-cores-and-climate-change/)

This is certainly a true statement, at least up to the speculative last sentence, but it also doesn't mean that changes in the concentration of atmospheric CO$_2$ *do influence* global temperatures. The claimed relationship is based on a theory as yet to be adequately examined with contemporary records.

There remains the question Gore ignored—if atmospheric CO$_2$ *is* the cause of the observed temperature changes in Figure 4-1, exactly what was causing atmospheric CO$_2$ to change? It certainly wasn't humans using fossil fuels.

For a more informed view of the authenticity of the 800-year lag:

Man-made global warming promoters claim the high correlation between carbon dioxide (CO$_2$) and atmospheric temperature (T) in the 420,000 year ice core record proves CO$_2$ causes T to change. Herein is demonstrated how the evidence conflicts with that belief…[54] [if you really want to know, *read referenced material*[54]]

Before moving on to examine the US government's recorded observations, the jury should ponder questions left unanswered by the prosecution:

- If changing CO_2 is such a powerful climate change force, how is it that during the 800 years of "lag," CO_2 is so easily dominated by natural forcings that the prosecution routinely discounts as an insignificant climate change force (Figure 3-3, page 41)?
- How is it possible during the Holocene interglacial (at extreme right of Figure 4-1) for the atmospheric CO_2 to grow dramatically while temperature drops off during the past 5,000 years?
- If changing CO_2 becomes the dominant climate change force *after* the initial 800 years of warming (as is often claimed), then *what is causing* atmospheric CO_2 to change *after* the initial 800 years?

To grasp how small 800 years is in context with the 800,000-year scale of Figure 4-1, a thin red vertical bar whose thickness approximates 800 years is shown in the gray area near the 275,000 on the "Age (years before present)" axis.

Can the prosecution identify *any consistent point during the ice age-interglacial cycles* in Figure 4-1 when natural climate change forces that brought the onset of an interglacial were overtaken by atmospheric CO_2 growth as the claimed cause of global climate change? No, it cannot. Frankly, the scale doesn't allow that fine a visual resolution of the data. If the prosecution is correct, where are the higher-resolution graphics to support its position? Can the prosecution identify *any consistent point* when natural climate change forces reasserted control of climate change as cooling set in that led to the end of an interglacial? Again, it cannot.

The prosecution's awkward attempt to explain away the 800-year CO_2 lag appears to be contrived. Furthermore, since climate cooling typically precedes atmospheric CO_2 decline, how did climate manage to cool after CO_2 had become the dominant climate change force?

Worth noting, the 100-ppm maximum range of CO_2 over the 800,000-year ice core record is not at all unusual in perspective with the geologic evidence-based range of atmospheric carbon dioxide over the past 550 million years (Figure 3-5, page 46).

While the evidence loosely links shorter-term ice age-interglacial cycles to Earth's orbital characteristics, the clear evidence from Antarctic ice core analysis:

- informs the jury that at the onset of an interglacial, atmospheric CO_2, lagging climate change by an average of 800 years, is clearly not the driving force behind climate change,
- strongly indicates that orbital linkage is *not the sole determinant* of the onset, duration, and termination of irregular ice age-interglacial cycles (Figure 4-1), and
- reveals no consistent evidence of any causal relationship that supports the prosecution's theory that changing atmospheric CO_2 drives climate change during ice age-interglacial cycles.

Near the end of this chapter, Greenland's GISP2 ice core record is examined to help the jury scrutinize the relationship between atmospheric CO_2 and climate during the 10,700 years of the current (Holocene) interglacial. As noted in Chapter 3, Earth is currently experiencing its *least typical* major climate regime, an interglacial between ice age cycles within an ice epoch of an ice era.

The prosecution's *inability to point to any evidence* over any meaningful period during the past 550 million years that consistently supports its theory that changing atmospheric carbon dioxide causes climate change is a major problem for the prosecution's climate change theory.[39]

Jurors interested in this debate over the 800-year lag should read both cited references[53–54] and compare the quality of the arguments.

While interesting, the jury is better served by leaving that debate for another time because the contemporary evidence over the past 140 years will compel the jury to reach the same conclusion that "the evidence conflicts with [climate change theory] belief," as explained in the second referenced quotation and its cited references.

Summary: The 800,000-Year Antarctic Ice Core Record:

1. There is no evidence in either the 550-million-year geologic or 800,000-year ice core record that consistently supports the prosecution's greenhouse gas climate change theory. Careful examination reveals both records contradict the prosecution's theory more often than they support it.
2. The 800,000-year ice core record shows that climate change preceded atmospheric CO_2 change by an average 800 years, directly contradicting the prosecution's climate change theory. The prosecution offers no plausible explanation for why it's claimed "weak" natural forcings dominate the theorized "strong" climate change force (changing atmospheric CO_2).

"Looking Out the Window" at the Recent Records

The US government maintains two pertinent historic records, one for annual global average atmospheric CO_2 and the other for annual global average surface temperature (GAST) over different surfaces (land, ocean, and merged land plus ocean). Temperature records are maintained as *anomalies* (the difference between the recorded temperature and a particular long-term average temperature).

The prosecution's greenhouse gas climate change theory alleges that changing (increasing amounts of) atmospheric CO_2 will cause corresponding global average surface temperature change (climate warming). For this reason, merged land plus ocean temperature records are used since they are the most representative measure available for global average surface temperature (GAST).

If the recorded evidence confirms atmospheric CO_2 changes *consistently* produce corresponding temperature changes (i.e., rising CO_2 typically causes warming, falling CO_2 typically causes cooling), then the two data records will necessarily be found to be highly correlated and supportive of climate change theory.

On the other hand, should recorded evidence reveal global atmospheric CO_2 growth is *uncorrelated* with global average temperature warming, then observed records will be seen to be in conflict with the prosecution's climate change theory that alleges *they must be* correlated (causation *creates* correlation). In that event, which do jurors suppose should be believed? Theory? Or the observed record?

The scientific method categorically invalidates any theory that fails to agree with the observed evidence in nature (in this case, observation in the form of recorded historic CO_2 and temperature measurements).

Is atmospheric CO_2 growth by any process, natural or human, a significant global climate change force?

The prosecution's greenhouse gas climate change theory[55–62] alleges it is, but what do the records reveal?

In his book *Global Warming—Myth or Reality? The Erring Ways of Climatology*,[63] a highly-regarded climatologist, the late Marcel Leroux, PhD (climatology) examines the evidence to identify significant climate change forces.

Leroux concludes (in Chapter 6, page 141):

Conclusion: The Greenhouse Effect is not the cause of climate change.

The possible causes, then, of climate change are:

- well-established orbital parameters on the palaeoclimatic scale, with climatic consequences slowed by the inertial effect of glacial accumulations;
- solar activity, thought by some to be responsible for half of the 0.6°C [1.1°F] rise in temperature, and by others to be responsible for all of it, which situation certainly calls for further analysis;
- volcanism and its associated aerosols (and especially sulphates), whose (short-term) effects are indubitable;
- and far at the rear, the greenhouse effect, and in particular that caused by water vapour, the extent of its influence being unknown.

That last bullet is very important because it indicates the following:

1. Climate change *is not* significantly (if at all) controlled by the greenhouse effect.
2. If the greenhouse effect has *any meaningful* climate change impact, that impact would be from the effects of atmospheric *water vapor, not carbon dioxide or methane!*

Doubly a PhD in climatology, Leroux expressed serious reservations about the validity of *any* greenhouse effect as a significant force for climate change.

In Chapter 5 of his book (at 5.5, page 100), he examines the question, "Is There Really A 'Greenhouse Effect'?" In that section, Leroux quotes *"Thieme (2002): 'the greenhouse gas hypothesis violates fundamentals of physics'"*[59] and observed, "Such a statement merits our attention."

Leroux continues:

As [Thieme's] support, the greenhouse effect scenario uses the examples of other planets, which serve as test beds to validate models. For example, Venus is considered to exhibit 'a quintessential greenhouse effect', with its surface temperature of 458°C, originally attributed to the carbon dioxide which represents 95% of its atmosphere (Courtin *et al.*, 1992). However, it seems that this temperature is the result of Venus' colossal atmospheric pressure, 92 times that of the mean pressure at the Earth's surface [combined with Venus' proximity to the Sun]. Now the infrared absorption of a gas increases with pressure, and so the resulting temperature depends mainly on pressure.

Thieme (2003) took this as his initial hypothesis when he expounded the following arguments:

.

.

.

3. If the radiation emitted by the Earth's surface were absorbed by the atmosphere, the absorbing air would warm up, and the initial structure of the air would be modified, most noticeably in its vertical temperature, density, and pressure profiles. Warm air tends to rise, following the basic principle of thermal convection: air warms up when in contact with the surface [conduction], and rises, transporting heat to upper levels. But air expands with altitude, and cools... The greenhouse effect (i.e., the return of heat downwards) does not occur; instead, updrafts are transformed into horizontal advections.

Then (at 5.6, page 100), in *The Greenhouse Effect "Cools" the Atmosphere*, Leroux discusses the findings of Lenoir (2001, *Climat de Panique*), who examines the climate impact of an atmosphere devoid of greenhouse gases, concluding that greenhouse gases contribute to atmospheric *cooling*. Leroux agrees, and so do many other scientists whose testimony the prosecution simply dismisses as the work of "denialists" while failing to address the substance of their testimony.

Yet as the jury will learn, it is not Leroux, Lenoir, Thieme, or any of the "human-caused global warming" skeptics who are in denial. Instead, it is the prosecution who denies the unequivocal testimony of Earth's long history of global average temperature and atmospheric CO_2. That history supports the views of Leroux, Lenoir, Thieme, et al., while invalidating the prosecution's climate change theory!

Leroux's conclusion is worth emphasizing (at 6.6, page 120):

> far at the rear, the greenhouse effect, and in particular that caused by water vapour, the extent of its influence being unknown…

On the back cover of his book, Leroux's parting advice:

> The most urgent priority for climatology… is to leave the IPCC in order that the discipline remains neutral and returns to the pursuit of its proper ends.

Those are prophetic words. If only Leroux's advice had been heeded. It is worth repeating:

> The most urgent priority for climatology…is to leave the IPCC…

For jurors who wish to learn more about the greenhouse effect from scientists who maintain a skeptical viewpoint consistent with the observed evidence, there are a number of worthy papers and books on the subject.[55–58, 64]

Defense testimony (*looking out the window*) in the form of real-world evidence is found in the annual records for global atmospheric CO_2 and temperatures. This evidence will be scrutinized to determine the extent to which observed *annual changes* in atmospheric CO_2 concentration show any meaningful relationship to observed *annual changes* in global average surface temperature (GAST).

US government records are kept as each year's temperature anomaly, which is the deviation of the actual measured temperature from a specific long-term average temperature.

Anomalies focus on year-to-year temperature *change* and can be quite helpful when jurors examine the evidence.

While record-keepers provide the timeframe used as the basis for the long-term average, the long-term average itself is rarely identified. Without knowing either the long-term average or one actual temperature measurement, it is impossible to reconstruct the actual *temperature* history upon which anomalies are based. For the jury's requirements, this is not a problem.

Since *temperature change* is more useful than actual temperature, the average temperature isn't needed. The temperature *change* between any two years is simply the difference between the temperature anomaly for each of those two years.

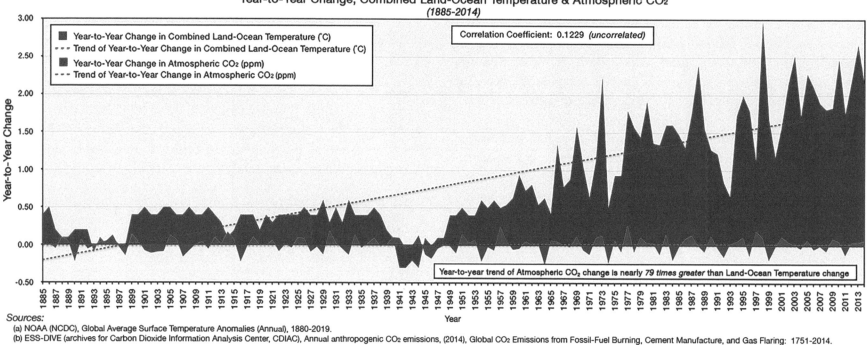

Figure 4-2
Year-to-Year Change, Combined Land-Ocean Temperature & Atmospheric CO₂
(1885-2014)

Sources:
(a) NOAA (NCDC), Global Average Surface Temperature Anomalies (Annual), 1880-2019.
(b) ESS-DIVE (archives for Carbon Dioxide Information Analysis Center, CDIAC), Annual anthropogenic CO₂ emissions, (2014), Global CO₂ Emissions from Fossil-Fuel Burning, Cement Manufacture, and Gas Flaring: 1751-2014.

The 130-Year Atmospheric CO_2 and Temperature Records (1885–2014)

With the aid of Figure 4-2 jurors can compare corresponding yearly atmospheric CO_2 change with global surface temperature change. Colored shading is used to assist the jury's scrutiny of important characteristics of these two key variables. Based on global records for yearly average atmospheric CO_2 and surface temperature anomalies[7, 65] annual atmospheric CO_2 change is shaded light green and corresponding annual merged land plus ocean temperature change (°C) is shaded cyan. Records are shown from 1885 through 2014. Since the correspondence of change is the key factor, the vertical scale is immaterial.

For their examination of contemporary climate records (late 19th century through today) the jury will primarily focus on the climate as represented by merged land plus ocean temperature records. These records provide the very best evidence for contemporary global climate change.

Note that temperatures above the zero of the year-to-year change scale represent warming, whereas those below the zero represent cooling below the long-term average.

Looking at Figure 4-2, in all but six years (4.6% of the 130 years), global average atmospheric CO_2 consistently grew. Annual atmospheric CO_2 growth rate after 1944 is dramatic, foretelling persistent rapid atmospheric CO_2 growth.

However, while atmospheric CO_2 was growing nearly every year, the magnitude of the growth varies significantly from year to year. This strongly suggests CO_2 growth is entirely divorced from yearly annual fossil fuel emissions (or, indeed, any relatively stable yearly anthropogenic emissions). As indicated by the growing light-green shaded areas, the record clearly shows that after 1944 the atmospheric CO_2 growth rate began a sudden much more rapid upward trend.

Ironically, during the first five years of the 1940s, global fossil fuel emissions increased nearly 50% (47.7%) over what they were during the prior five years of the late 1930s, yet global average CO_2 *declined* for four of the first five years of the 1940s (light-green shading in negative area).

This calls into question the prosecution's oft-repeated allegation that fossil fuel emissions are *primarily responsible* for the growth of atmospheric CO_2 when the evidence clearly indicates the contrary. If, as alleged by the prosecution, fossil fuel emissions were responsible for the growth of atmospheric CO_2, how is it possible for atmospheric CO_2 to have *declined for four straight years* in the early 1940s when the first five years of the 1940s experienced higher annual fossil fuel emissions than *any* prior five-year period since fossil-fuel emissions records began in the 19th century?

It is somewhat ironic that while the four years from 1941 through 1944 produced the only noticeable time during the 20th century when atmospheric CO_2 actually *declined* (-0.80 ppm), in defiance of the prosecution's climate change theory global average temperatures warmed +0.095°C (+0.171°F).

An even more persistent clear contradiction to the prosecution's theory is the consistent variation of *annual* temperature change between warming (above the zero) and cooling (below the zero) of the long-term average throughout the 130-year record. The trend of temperature change (dashed blue line) is so small it is virtually indistinguish-

able from the zero of the year-to-year change axis (left side of the graph). The lack of any corresponding temperature reaction to the strong mid-century atmospheric CO_2 growth offers jurors compelling evidence in stark contradiction to the prosecution's alleged relationship between growing atmospheric CO_2 and climate warming.

The dominance of green-shaded areas above the zero of annual atmospheric CO_2 change tells the jury that atmospheric CO_2 increased to some extent nearly every year since 1885. Even the more modest growth rates prior to 1940 reveal a persistent small annual atmospheric CO_2 growth. Nevertheless, global average temperature shows no corresponding reaction throughout the period 1885–2014. Jurors should note the sharp contrast between modest temperature variations and the dramatic change in the rates of atmospheric CO_2 growth over this 130-year period.

Keeping in mind the prosecution's climate change theory, do these records reveal any consistent relationship in the 130-year record (1885–2014) that even suggests global atmospheric CO_2 change causes corresponding global average surface temperature (GAST) change? No, they do not.

According to theory, such a persistent CO_2 growth should be accompanied by a persistent temperature warming, yet the cyan-shaded temperature changes show no evidence of any corresponding temperature warming beyond a trivial level consistent with post-Little Ice Age climate recovery and too small to discern in Figure 4-2's 130-year view.

This is clear evidence the prosecution's theory is not supported by the observed behavior of changing atmospheric CO_2 and temperature.

For more than 70 years (after 1940), the growth rate of atmospheric CO_2 is *dramatically higher* than the much lower growth rate for global average surface temperatures.

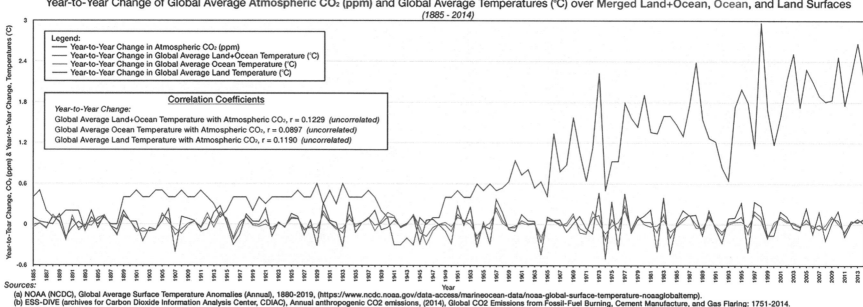

Figure 4-3

Year-to-Year Change of Global Average Atmospheric CO₂ (ppm) and Global Average Temperatures (°C) over Merged Land+Ocean, Ocean, and Land Surfaces
(1885 - 2014)

Sources:
(a) NOAA (NCDC), Global Average Surface Temperature Anomalies (Annual), 1880-2019, (https://www.ncdc.noaa.gov/data-access/marineocean-data/noaa-global-surface-temperature-noaaglobaltemp).
(b) ESS-DIVE (archives for Carbon Dioxide Information Analysis Center, CDIAC), Annual anthropogenic CO2 emissions, (2014), Global CO2 Emissions from Fossil-Fuel Burning, Cement Manufacture, and Gas Flaring: 1751-2014.

Because the relationship between atmospheric CO₂ *change* and temperature *change before* 1944 is radically inconsistent with the relationship *after* 1944, the jury will also examine the 75-year record from 1940 to 2014. This disconnect with the prosecution's theory will be explored in greater detail later in the section *The 139 Years of Atmospheric CO₂ and Temperature Changes (1880–2018)*.

If atmospheric CO₂ *change* is a major force driving temperature *change*, how is it possible for atmospheric CO₂ change to be so radically different before and after 1940 while temperature change remains fairly consistent throughout the 130 years? Were the climate change theory valid, that would simply not be possible in the absence of some other stronger mitigating climate change force. The laws of physics did not change after 1940.

Figure 4-3 shows both year-to-year atmospheric CO₂ and year-to-year temperature change profiles for all three surface regions for which temperatures are recorded (land, ocean, merged land plus ocean). Curves are unshaded so the jury can view all three temperature sets on the same chart.

Plotted values dipping into the shaded area below the zero of the vertical axis show temperature fell (or CO₂ declined) below the long-term average. When the plotted lines descend, the annual rate of change declined; when they ascend, the annual rate of change increased.

Contrary to the prosecution's theory-based claim that observed 20th-century temperature change is a reaction to atmospheric CO₂ growth alleged to be caused by fossil fuels, the relationships revealed in Figure 4-3 offer compelling evidence *no corresponding relationship exists* between annual global atmospheric CO₂ change and annual

GAST change (regardless of surface region, merged land plus ocean, land, or ocean)!

During the period 1940 to 2014, a significant year-to-year growth of atmospheric CO_2 (red line) is not reflected in any GAST change. While GAST changes very little throughout the 130 years to 2014, annual CO_2 growth dramatically increases after 1944. This behavior is inconsistent with the relatively small incremental growth of yearly fossil fuel emissions, clearly demonstrating to the jury that the dramatic growth of atmospheric CO_2 cannot be attributed to fossil fuel use.

If the prosecution's greenhouse gas climate change theory were valid, then after 1944, year-to-year GAST change should consistently be positive and dramatically increasing, matching the characteristics of rapidly growing atmospheric CO_2. But it isn't.

Included with Figure 4-3 are correlation coefficients between atmospheric CO_2 *change* and GAST *change* for each temperature region. Ranging from 0.09 to 0.12, correlation coefficients provide *compelling confirmation of the lack of any relationship* between changing atmospheric CO_2 and climate change. What clearer evidence could the jury view to inform them the prosecution's greenhouse gas climate change theory is invalid than *the lack of any correlation* between changing atmospheric CO_2 and changing temperature?

While record sets can be correlated yet not have a causal relationship between each other, whenever two data sets are *uncorrelated, there cannot possibly be any causal relationship* between the two. Therefore, during the period 1885 to 2014, it is simply not possible for changes in atmospheric CO_2 to have caused changes in global average temperature regardless of the source of any observed atmospheric CO_2 growth.

By marking those years when both atmospheric CO_2 and GAST change in the same direction (both up, in green or both down, in red), Figure 4-4 visually illustrates the dramatic lack of any correlation between yearly atmospheric CO_2 change and yearly temperature change from 1885 through 2014.

While bar *direction* (up or down) is significant, bar length has no meaning. For each year when CO_2 and temperature change move in the same direction (both up, both down, both stable) a bar above the horizontal baseline is displayed. This indicates consistency with the prosecution's climate change theory. This comparison offers the jury compelling visual evidence confirming that atmospheric CO_2 growth is not driving global average temperature change.

The double-bar chart style of Figure 4-4 is specifically created to help the jury visually confirm that the prosecution's theorized causal relationship between annual atmospheric CO_2 change and annual global average surface temperature change simply does not exist. If the prosecution's theory had any merit, the vast majority of the bars would be green (above the line).

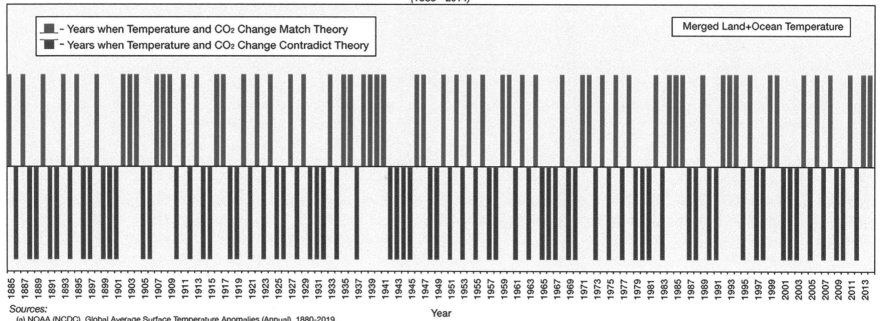

Figure 4-4
Does *Direction* of Annual CO₂ Change Match *Direction* of Annual Temperature Change?
(1885 - 2014)

■ - Years when Temperature and CO₂ Change Match Theory
▮ - Years when Temperature and CO₂ Change Contradict Theory

Merged Land+Ocean Temperature

Year

Sources:
(a) NOAA (NCDC), Global Average Surface Temperature Anomalies (Annual), 1880-2019.
(b) ESS-DIVE (archives for Carbon Dioxide Information Analysis Center, CDIAC), Annual anthropogenic CO₂ emissions, (2014), Global CO₂ Emissions from Fossil-Fuel Burning, Cement Manufacture, and Gas Flaring: 1751-2014.

The history of atmospheric CO_2 change and temperature change used to create Figure 4-4 is based on NOAA's NCDC (National Climate Data Center) data (maintained by ESRL) for annual global average atmospheric CO_2 records and annual global average surface temperature (GAST) anomalies for merged land plus ocean (global) temperatures spanning 130 years from 1885 to 2014.[1, 66]

Because greenhouse gases react to limited wavelengths of (infrared, IR) radiant heat, any theorized consequent surface temperature change must be virtually instantaneous (since radiant heat transfer occurs at the speed of light). Consequently, if the prosecution's greenhouse gas climate change theory were valid, then absent a strong mitigating opposing force, a full year of atmospheric CO_2 change should certainly produce evidence of a corresponding detectable temperature change.

But it doesn't. In stark contradiction to climate change theory, from 1944 to 1974 global climate cooled -0.361°C (-0.650°F). Yet throughout that period atmospheric CO_2 was rapidly growing while global temperatures were cooling.

If the prosecution's climate change theory were valid, real-world compliance with the greenhouse gas theory would certainly be evident over three full decades. Yet the only consistency observed is the records' contradiction to climate change theory.

Emphasizing that contradiction, Figure 4-4's bars are about evenly split between downward and upward, suggesting a chance relationship is the *only* relationship that exists between changing atmospheric CO_2 and temperature change.

This evidence alone constitutes invalidation of the prosecution's theory by *the scientific method* for its failure to comply with observations (atmospheric CO_2 and temperature records).

Figure 4-5
Percentage of Time When
Atmospheric CO2 and Global Average Temperature
Year-to-Year Changes Conform with Theory
(1885-2014)

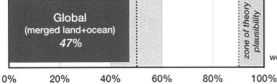

Figure 4-5 is based on the temperature records for merged land plus ocean surfaces as most representative of global average temperature. Note that the percentage of years the observed record agrees with theory is just 47%. If the prosecution's theory were valid, it would conform with the record at least 90% of the time, and the green horizontal bar would be in the gray plausibility zone (extreme right, between 90% and 100% agreement with theory).

Ironically, the actual relationship between climate change and atmospheric CO_2 change is in the region of pure chance and can be simulated by the mere flip of a coin! Imagine, a simple flip of the coin does a better job predicting the relationship between changing atmospheric CO_2 and temperature change than the costly CMIP5 array of climate simulation models based on the prosecution's dubious climate change theory!

Based on Figure 4-4, Figure 4-5 makes several interesting points for the jury:

- The agreement of total atmospheric carbon dioxide change and global surface temperature change with greenhouse gas climate change theory is just 47% of the years between 1885 and 2014. That's not agreement with theory; it's invalidation.
- There is no evidence that temperature change corresponds with CO_2 change for any reason other than pure chance (a coin flip).

There is no valid reason to continue to tweak a costly array of the prosecution's CMIP5 climate simulation models that (as should be expected) *relentlessly* over-predict temperature because they are based on a flawed theory that incorrectly chains changing temperature to changing atmospheric CO_2 despite no evidence of any such relationship existing in any record for atmospheric CO_2 change and temperature change.

The evidence fully supports the conclusions of the late Dr. Marcel Leroux in his text on climate change (*Global Warming—Myth or Reality? The Erring Ways of Climatology*[63]) wherein he assesses the forces that influence climate change:

> far at the rear, the greenhouse effect, and in particular that caused by water vapour, the extent of its influence being unknown.

Leroux understood atmospheric CO_2 has no capacity to affect climate change. The clear evidence confirms that the "trace gas" carbon dioxide (CO_2) has no discernible impact on climate change.

While the distinct change in the rate of annual (year-to-year) atmospheric CO_2 growth in the vicinity of 1940 is quite evident, the source of that change is unclear.

The dramatic increase in the annual growth of atmospheric CO_2 after 1940 had no discernible impact on annual temperature change.

Summarizing the 130-Year Record (1885–2014)

1. 130 years of global averages for surface temperature and atmospheric CO_2 offers jurors compelling evidence the prosecution's greenhouse gas climate change theory is contradicted by real-world evidence.
2. The scientific method requires theory to conform with real-world evidence. There is ample nonconformity to unequivocally conclude the greenhouse gas climate change theory is invalid.
3. Annual global average atmospheric CO_2 change is strongly uncorrelated with annual global average temperature change. No correlation means no causation is possible.
4. The flip of a coin is a far more accurate predictor of the response of annual temperature change to annual atmospheric CO_2 change than the prosecution's costly array of theory-based climate simulation models.
5. During the first five years of the 1940s, global fossil fuel emissions increased nearly 47.7% over what they had been during the last five years of the 1930s. At the time, this was the highest fossil-fuel emissions increase for any five-year period since emissions records began in the mid-19th century. Over that same five-year period, global atmospheric CO_2 recorded its most pronounced *decline* over any five-year period from 1885 to the present. This constitutes clear evidence the prosecution has no real grasp of the extent to which annual fossil fuel emissions accumulate in the atmosphere and contribute to CO_2 growth.
6. The evidence convincingly contradicts the prosecution's assertion that fossil fuel emissions are the primary source of atmospheric CO_2 growth.

The 75-Year Atmospheric CO_2 and Temperature Records (1940–2014)

The persistent dramatic increase in the rate of atmospheric CO_2 growth after 1944 suggests the jury should examine the 75-year-record between 1940 and 2014, a period during which the prosecution maintains there is ample evidence supporting its climate change theory.

The 75 years between 1940 and 2014 (Figure 4-6) show no causal relationship between a persistent growth of atmospheric CO_2 (red) every year after 1944 and the recorded annual warming and cooling of global surface temperatures (cyan, brown, and blue for merged land plus ocean, land, and ocean, respectively).

As was observed for the 130-year record, there isn't even *a hint* of a linkage between global atmospheric CO_2 change and global surface temperature change (the light blue-gray background indicates a change below the long-term average). Note the period 1941–1944 is the *only period* during which atmospheric CO_2 is observed to *decline* below the long-term average (red curve in negative territory) while, as noted previously, there was a general *increase* in global average temperatures (curves in positive territory). Further, for several years after 1944, CO_2 dramatically *increased* while temperatures *decreased* below their long-term average. These observations are a clear contradiction to the prosecution's theory that claims global atmospheric CO_2 change causes corresponding global temperature change.

Figure 4-6
Global Average Atmospheric CO₂ (ppm) and Global Average (°C) for Combined Land-Ocean, Ocean, and Land Surfaces
(1940 - 2014)

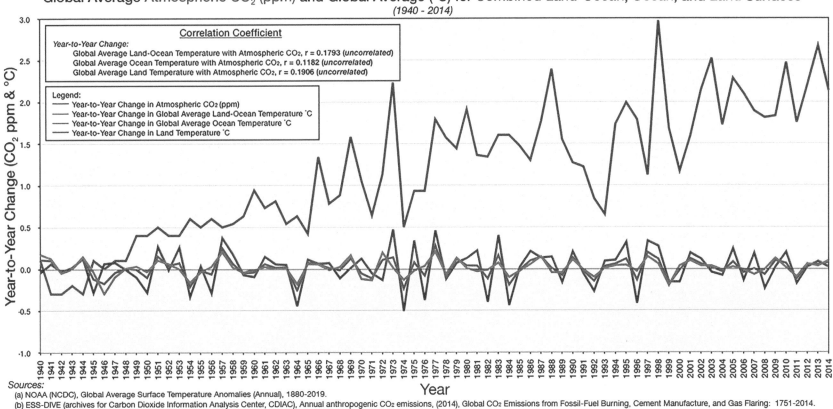

Sources:
(a) NOAA (NCDC), Global Average Surface Temperature Anomalies (Annual), 1880-2019.
(b) ESS-DIVE (archives for Carbon Dioxide Information Analysis Center, CDIAC), Annual anthropogenic CO₂ emissions, (2014), Global CO₂ Emissions from Fossil-Fuel Burning, Cement Manufacture, and Gas Flaring: 1751-2014.

The observed lack of any evidence of a causal relationship between atmospheric CO_2 change and temperature change is confirmed by the extremely low correlation coefficients between atmospheric CO_2 and temperature change (ranging from 0.12 to 0.19).

Despite the clear record showing atmospheric CO_2 growing dramatically each year after 1944, global average surface temperature records contain no evidence of any corresponding warming.

Based on these 75 years (1940-2014), a double-bar chart similar to that of Figure 4-4 is created as Figure 4-7. This chart displays the history of the annual change of global average atmospheric CO_2 and global average surface temperature in terms of whether the changes support the prosecution's greenhouse gas climate change theory (green bars) or whether they contradict theory (red bars).

Figure 4-7
Does *Direction* of Annual CO₂ Change Match *Direction* of Annual Temperature Change?
(1940 - 2014)

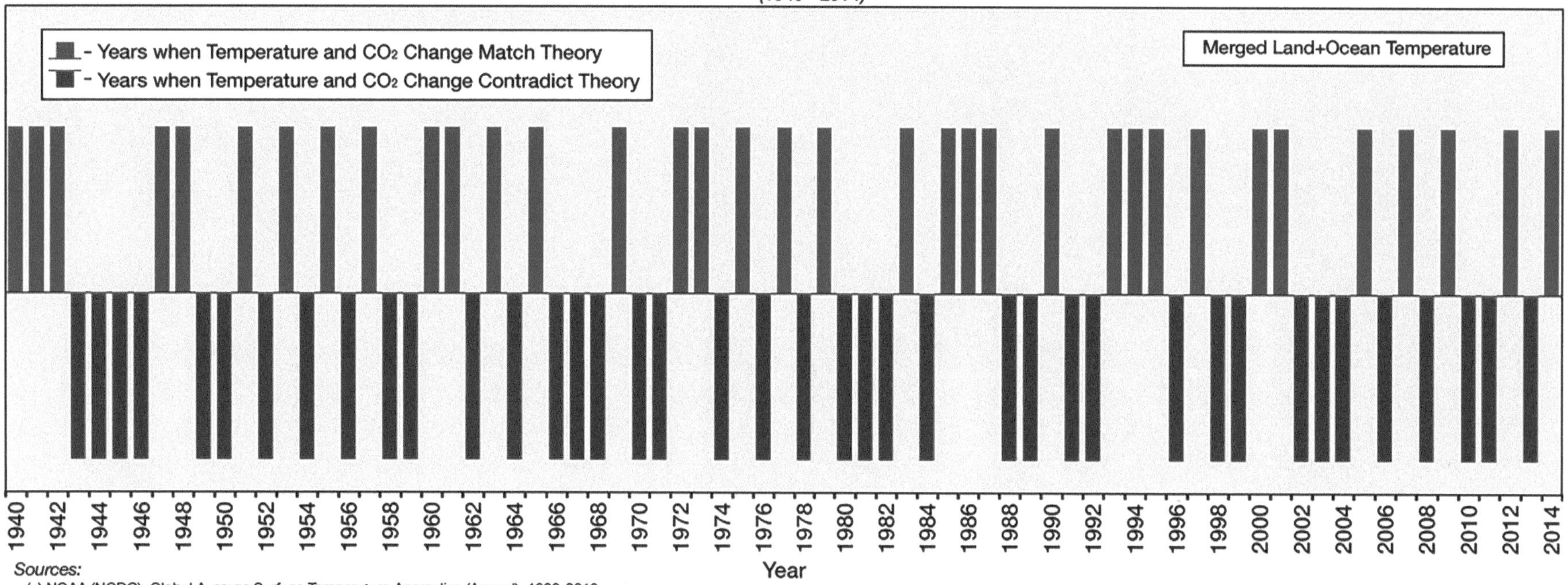

■ - Years when Temperature and CO₂ Change Match Theory
▼ - Years when Temperature and CO₂ Change Contradict Theory

Merged Land+Ocean Temperature

Year

Sources:
(a) NOAA (NCDC), Global Average Surface Temperature Anomalies (Annual), 1880-2019.
(b) ESS-DIVE (archives for Carbon Dioxide Information Analysis Center, CDIAC), Annual anthropogenic CO₂ emissions, (2014), Global CO₂ Emissions from Fossil-Fuel Burning, Cement Manufacture, and Gas Flaring: 1751-2014.

Ironically, just as pronounced as with the 130-year record, there are fewer upward bars (*47% supporting theory*) than downward bars (*53% contradicting theory*). This chart reaffirms that over the 75 years from 1940 to 2014 no consistent relationship exists between global average CO₂ change and global average temperature change to support the prosecution's greenhouse gas climate change theory.

This is particularly informative since the prosecution maintains its theory is more strongly evident in more recent years, whereas the testimony of the record clearly tells the jury the prosecution's theory is strongly contradicted during the years 1940 to 2014.

Recall the "if this, then that" nature of theory. For any theory to be valid, the observed evidence should consistently support this theorized relationship. If the prosecution's theory were valid, then virtually every bar would be above the horizontal line. But that is clearly not the testimony of this evidence.

The theorized radiation-based greenhouse gas warming over the course of a year's average should be readily apparent, persistent, strong, and undeniable. But that is not the case. There isn't a scintilla of real-world evidence to support the prosecution's climate change theory.

Whenever a theory contradicts observation, the scientific method confidently affirms that theory is invalid. So should the jury.

As with Figure 4-5, Figure 4-8 shows that during the 75 years of dramatic atmospheric CO_2 growth, real-world evidence agrees with theory just 47% of the time!

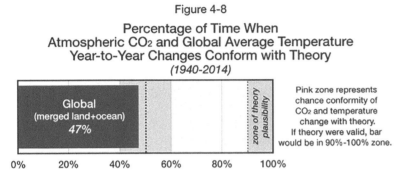

Figure 4-8

**Percentage of Time When
Atmospheric CO_2 and Global Average Temperature
Year-to-Year Changes Conform with Theory**
(1940-2014)

Figure 4-8 makes several compelling points:

- Real-world records for global averages of atmospheric carbon dioxide and surface temperatures are more likely to contradict than conform with greenhouse gas climate change theory. Theory conformity occurs in just 47% of the 75 years from 1940 through 2014. To be valid, the theory should consistently conform with observations at least 90% of the time (with good explanation for the nonconformity of the other 10% of the time).
- There is no evidence that temperature change corresponds with CO_2 change for any reason other than pure chance.

In other words, *real-world evidence* reveals a fifty-fifty coin flip would be a far more accurate climate change predictor than an invalid greenhouse gas theory and the theory-based CMIP5 array of climate simulation models.

Basing its charges on an unproven theory without first confirming its theory with evidence, the prosecution acted hastily when it indicted defendant carbon dioxide as a climate change force. Evidently, the

prosecution thought it could take advantage of a brief chance correspondence between growing atmospheric CO_2 and warming temperatures to assert its flawed theory.

In fact, the prosecution's theory asserts a relationship between changing CO_2 and changing temperature that is contradicted by real-world evidence covering every meaningful time period. The observed chance relationship between changing atmospheric CO_2 and global temperature change completely invalidates the prosecution's theory.

Recall that the 75 years from 1941 to 2014 included the beginning of the rapid increase in the rate of atmospheric CO_2 growth after 1944.

The 1941 to 1974 period of strong global cooling covers the only years of the 20th century during which atmospheric CO_2 declined (1941–1944).

Figure 4-9 shows the dramatic contradiction to the prosecution's climate change theory during the period 1941 to 1974. From 1941 to 1944, while atmospheric CO_2 declined by nearly 1 ppm, *GAST warmed* by nearly 0.1°C; from 1944 to 1974, while atmospheric CO_2 dramatically grew by 20 ppm, *GAST cooled* by -0.36°C. In other words, when atmospheric CO_2 fell, temperatures warmed; when atmospheric CO_2 grew, temperatures cooled, *just the opposite* of the prosecution's theory-based relationship!

US government records unambiguously show no theory-consistent relationship exists between atmospheric CO_2 change and global temperature change during both the 130 years from 1885 through 2014 and the 75 years from 1940 through 2014.

From 1941 through 1974, a particularly strong contradiction to the prosecution's theory is strikingly apparent. These records clearly behaved no differently than had humans never used fossil fuels.

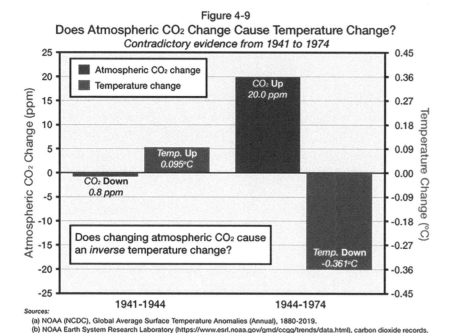

Figure 4-9
Does Atmospheric CO₂ Change Cause Temperature Change?
Contradictory evidence from 1941 to 1974

Sources:
(a) NOAA (NCDC), Global Average Surface Temperature Anomalies (Annual), 1880-2019.
(b) NOAA Earth System Research Laboratory (https://www.esrl.noaa.gov/gmd/ccgg/trends/data.html), carbon dioxide records.

No evidence in the observed record supports either greenhouse gas climate change theory or any human causation related to climate change.

Summary: The 75-year record (1940–2014)

1. Seventy-five years of global averages for surface temperature and atmospheric CO_2 provide clear evidence contradicting the prosecution's greenhouse gas climate change theory.
2. During the 34-year period from 1941 through 1974, a strong contradiction to climate change theory is clearly evident in the records. When atmospheric CO_2 declined, temperatures warmed; when atmospheric CO_2 grew, temperatures cooled!
3. In accordance with *the scientific method* that requires every theory to conform with real-world evidence, there is ample compelling evidence of nonconformity to unequivocally invalidate the greenhouse gas climate change theory.
4. Annual global average atmospheric CO_2 change is strongly uncorrelated with annual global average temperature change. No correlation means no causation is possible.
5. The flip of a coin is a far more accurate predictor of climate change than the prosecution's climate simulation models.

Before examining the 60-year period (1951 through 2010) that the prosecution touts as providing definitive evidence supporting its theory, the strong mid-20th-century change in the growth of atmospheric CO_2 suggests an excellent opportunity for the jury to peek at some additional evidence covering the full range of 139 years, 1880 through 2018.

The 139 Years of Atmospheric CO_2 and Temperature Changes (1880–2018)

Jurors have heard prosecution agents claim there is "a 99.9999 percent chance that humans are the cause of global warming."[67] Such sophistry is the product of the prosecution's theory-based studies that fail to account for the most important factor—real world evidence.

Forget theory, what does the *real-world record* plainly show?

Table 4-1 is based on the 139-year record of atmospheric CO_2 and global average surface temperatures (merged land plus ocean). The 139 years are broken into two parts:

1. the 65 years from 1880 through 1944, *prior* to the rapid increase in atmospheric CO_2, and
2. the 74 years from 1945 through 2018, *after* the rapid growth of atmospheric CO_2 began.

The annual rate of warming averaged 0.0065°C per year during the first 65 years and virtually identical at 0.0068°C per year during the remaining 74 years. The difference (less than 0.0004°C per year) is much smaller than the recording accuracy of the records! The 139-year average annual GAST change of +0.0066°C per year reflects the net 139-year change of 0.92°C (1.66°F). These averages testify that, unlike the annual rate of change observed for atmospheric CO_2, there is little difference in the annual rate of warming throughout the 139 years examined.

Table 4-1
139 Years of Compelling Evidence: Atmospheric CO_2 Change *Is Not* A Climate Change Force
(1880 through 2018)

Period	Years	ΔT (°C)	ΔT (°C)/yr	ΔCO2	ΔCO2/yr
1880 - 1944	65	0.42	0.0065	19.40	0.2985
1945 - 2018	74	0.50	0.0068	98.32	1.3286
1880 - 2018	139	0.92	0.0066	117.72	0.8469

Sources: CDIAC & MLO CO2 records and NOAA/NCDC GAST (land+ocean merged) at
https://www.ncdc.noaa.gov/data-access/marineocean-data/noaa-global-surface-temperature-noaaglobaltemp

The annual rate of CO_2 growth during the 74-year period following the onset of rapid CO_2 growth was 445% *higher than* the rate of CO_2 growth during the first 65 years of the 139 years examined! According to the prosecution's climate change theory, this rapid rate of CO_2 growth should have been easily observed in the rate of temperature warming over the same recent 74-year period. But it wasn't.

To summarize, while the rate of warming was virtually unchanged for 139 years, the rate of atmospheric CO_2 growth was more than four times greater during the last 74 years during which *there was no corresponding increase* in the rate of warming. This observed record offers jurors a stunning rejection of climate change theory. This example is also described in Defense Exhibit D-5.

If the prosecution's climate change theory claiming CO_2 growth causes climate warming were valid, a pronounced increase in the rate of warming during the period of rapid CO_2 growth would be evident. But there is *virtually no change* in the rate of warming (an average 0.0003°C per year, a rate that, if persistent, would produce a 1,000-year global temperature increase of just 0.3°C).

The rate of temperature warming being *completely independent* of the rate of atmospheric CO_2 growth throughout the 139 years from 1880 through 2018 is further compelling evidence the prosecution's theory is contradicted by 139 years of real-world evidence (from 1880 through 2018).

The prosecution would like the jury to look only at the overall average changes and deduce growing atmospheric CO_2 causes warming temperatures. Why? Because the actual evidence contradicts the prosecution's alleged relationship.

Figure 4-10 graphically illustrates the evidence in Table 4-1 that temperature warming is *independent of* the growth of atmospheric CO_2.

The only reasonable conclusion the jury can draw from the clear testimony of this real-world evidence is that the prosecution's allegations against both atmospheric CO_2 and fossil fuel use are unsupportable. This realization nullifies any claim that humans bear any responsibility *for any portion of recent observed routine climate warming!*

Do any of these revelations suggest the prosecution's allegation of 99.9999% certainty is anything other than pure fantasy designed to deceive the jury?

Studies performed on the basis of the prosecution's invalid theory are as worthless as the complex yet poorly-performing CMIP5 array of climate simulation models. There can be no other conclusion from the testimony of the past 139 years of GAST and atmospheric CO_2 records examined.

Figure 4-10

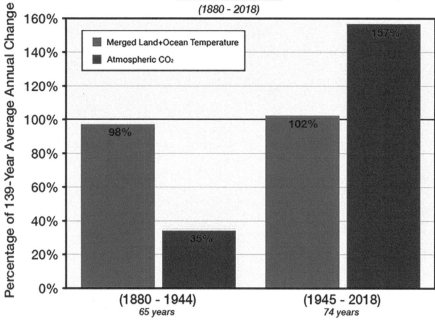

139 Years of Atmospheric CO₂ and Temperature Change Prove Human's Are _Not Responsible_ for Climate Change
(1880 - 2018)

Sources: CDIAC & MLO CO₂ records and NOAA/NCDC GAST (land+ocean merged) at
https://www.ncdc.noaa.gov/data-access/marineocean-data/noaa-global-surface-temperature-noaaglobaltemp

This record is consistent with and supports the conclusions drawn from the 130-year and 75-year records examined earlier. It is also consistent with the 550-million-year geologic record.

Correlation coefficients and the double-bar charts further reaffirm the testimony of each of these views of the evidence obtained by simply *looking out the window* at real-world records.

Summary: The 139-Year Record (1880–2018)

1. The 139 years of global averages for surface temperature and atmospheric CO_2 provide further compelling evidence con-

tradicting the prosecution's greenhouse gas climate change theory.

2. In accordance with *the scientific method* that requires every theory to conform with real-world evidence, there is ample evidence of nonconformity to unequivocally declare the greenhouse gas climate change theory invalid.

3. Annual global average atmospheric CO_2 change is dramatically greater after 1944 than it was prior to 1945, yet the rate of global average temperature growth remains virtually unchanged and trivially small throughout the 139-year period. Such real-world behavior is highly inconsistent with the prosecution's climate change theory.

4. The prosecution would do well to investigate alternate forces at work affecting Earth's climate and abandon its fixation on innocent fossil fuel emissions and atmospheric carbon dioxide, a gas that has no impact on global climate change and is the gas most essential to all life on Earth.

The 64 Years of Atmospheric CO_2 and Temperature Changes (1912–1976)

It is generally acknowledged that thirty years of observed records is sufficient time to constitute evidence of climate change. The jury has already discovered that prior to 1945 atmospheric CO_2 was growing at a slow pace consistent with what might be expected as slowly warming ocean temperatures emitted relatively more CO_2 to the atmosphere following the Little Ice Age.

After 1944, atmospheric CO_2 began to grow dramatically, far outpacing the pre-1945 growth rate. The prosecution would have the jury believe the dramatic increase of atmospheric CO_2 growth was dominated by the atmospheric accumulation of fossil fuel CO_2 emissions that are claimed to be responsible for changing climate.

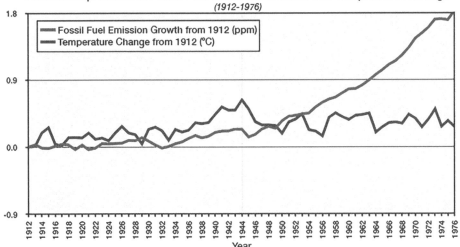

Figure 4-11
Relationship Between Fossil Fuel Emission Growth and Temperature Change
(1912-1976)

Data source: NOAA/NCDC global average yearly fossil fuel emissions, land+ocean GAST anomalies

Note that the fossil fuel emissions contribution to total atmospheric CO_2 is *significantly less than* the growth of fossil fuel emissions owing to the reabsorption of a substantial portion of fossil fuel emissions by Earth's oceans and its biosphere. The prosecution alleges human activity's fossil fuel emissions are the dominant source of atmospheric CO_2 growth. Yet both the evidence and many highly-qualified expert scientists challenge that prosecution theory, contending fossil fuel emissions are responsible for at most just 3% to 4% of atmospheric CO_2 growth. Figures 4-11 through 4-13 reveal the dubious nature of the prosecution's theory and offer the jury compelling evidence that fossil fuel emissions are a highly unlikely candidate for having any detectable impact on climate because there is no evidence that growing atmospheric CO_2 from *any* source has any impact on either temperature change or global climate.

What else do these graphs tell the jury?

But what does the evidence testify to? The sixty-four years, 1912-1976, span the hot decade of the 1930s as well as the multi-decade global cooling that began in the early 1940s and continued into the late 1970s.

What do historic records reveal about global fossil fuel emissions during those same sixty-four years?

Figure 4-11 together with Figures 4-12 and 4-13 (next page) present visual evidence covering the sixty-four years, spanning 1912–1976, and its two thirty-two-year components, 1912–1944 and 1944–1976. These graphs expose the true (chance) relationship between fossil fuel emissions growth and global average surface temperature change. These charts show the growth and decline from the chart's starting date for both fossil fuel emissions and global average surface temperature (GAST) for the entire 64-year range and its two component 32-year ranges.

Covering the sixty-four years, 1912-1944, Figure 4-12 shows the jury that up until the early 1940s, both fossil fuel emissions and temperature were gradually increasing. This suggests the two *might be* correlated, but it does not suggest which might be driving the other or whether any causal relationship between the two actually exists.

However, from the mid-1940s through 1976 (Figure 4-13), the records produce a *negative* correlation (-0.046) as fossil fuel emissions dramatically *increase* while temperature trends slowly *downward* through the multi-decade cooling from the 1940s into the late 1970s that led to fears a new ice age might be starting!

Clearly, the prosecution's theorized relationship between fossil fuel emissions and temperature (climate change) is non-existent during those sixty-four years. If the theorized relationship doesn't exist over a span of sixty-four years, shouldn't the prosecution be required to explain why?

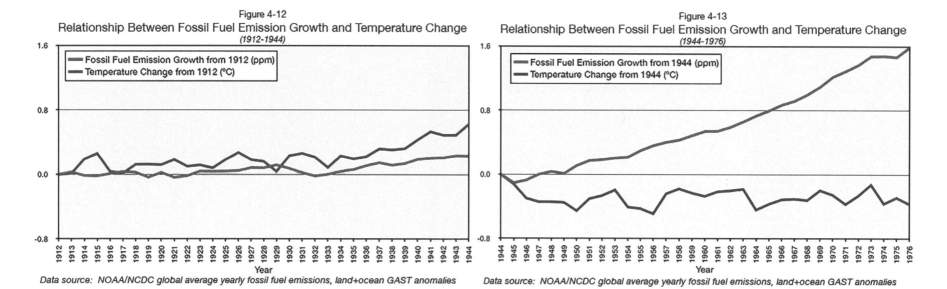

Figure 4-12
Relationship Between Fossil Fuel Emission Growth and Temperature Change
(1912-1944)

Figure 4-13
Relationship Between Fossil Fuel Emission Growth and Temperature Change
(1944-1976)

Data source: NOAA/NCDC global average yearly fossil fuel emissions, land+ocean GAST anomalies

Data source: NOAA/NCDC global average yearly fossil fuel emissions, land+ocean GAST anomalies

Figures 4-12 and 4-13 offer the jury clear evidence no consistent relationship between changing fossil fuel emissions and changing temperature exists. Covering the initial thirty-two years from 1912-1944, Figure 4-12 shows slight growth for both fossil fuel emissions and temperature. Yet no consistent relationship is evident.

NOAA's records for total atmospheric CO_2 over those same sixty-four years provide the *coup de grâce* for the prosecution's greenhouse gas climate change theory.

During the thirty-two years, 1912-1944, NOAA records show global average atmospheric CO_2 grew by 9.2 ppm while merged land plus ocean global average surface temperatures *warmed* by 0.55°C (nearly 1°F). Yet during the following thirty-two years from 1944 through 1976, NOAA's records show global average atmospheric CO_2 grew by 21.8 ppm as GAST *cooled* by -0.28°C (-0.51°F).

Got that? For thirty-two years Earth's climate *sharply warmed* during very modest atmospheric CO_2 growth. This was followed by thirty-two

years during which Earth's climate *dramatically cooled* by 0.51°F as atmospheric CO_2 grew 2.37 times what it had grown during the previous thirty-two years of sharp warming!

This sixty-four year record constitutes compelling evidence that something is very wrong with the prosecution's greenhouse gas climate change theory that strongly relates climate change to atmospheric CO_2 change when, in fact, the recorded evidence strongly *contradicts* that relationship.

The 60-Year Atmospheric CO_2 and Temperature Records (1951–2010)

The prosecution cites 1951 to 2010 as a 60-year period providing *solid evidence* atmospheric CO_2 growth, claimed to be primarily caused by fossil fuel CO_2 emissions, is warming the climate:

> It is *extremely likely* that more than half of the observed increase in global average surface tempera-

ture from 1951 to 2010 was caused by the anthropogenic increase in GHG [greenhouse gas] concentrations and other anthropogenic forcings together.[45]

Extremely likely? Based on what evidence? The prosecution's own theory and its theory-based simulation models? Legitimate science does not use a theory to validate itself.

Clearly, if records for *total* atmospheric CO_2 do not support the prosecution's theory, then the prosecution's cited allegation against any anthropogenic emissions of CO_2 is entirely unsupportable, and the size of the anthropogenic contribution to total atmospheric CO_2 is completely irrelevant.

Does the testimony of the 60-year record from 1951 to 2010 support the prosecution's bold claim?

Figure 4-14 shows the yearly rate of atmospheric CO_2 growth (red) is consistently positive every year across the 60 years from 1951 through 2010. At the same time, annual temperature change consistently varies above and below the zero axis around the long-term average over all surfaces (land, ocean, and merged land plus ocean). As atmospheric CO_2 persistently builds up, no related temperature change is observed.

Correlation coefficients between year-to-year atmospheric CO_2 change and temperature change testify to the *impossibility* of any causal relationship between the two. Note the many years when temperatures dropped while CO_2 increased, just the opposite of the prosecution's theory.

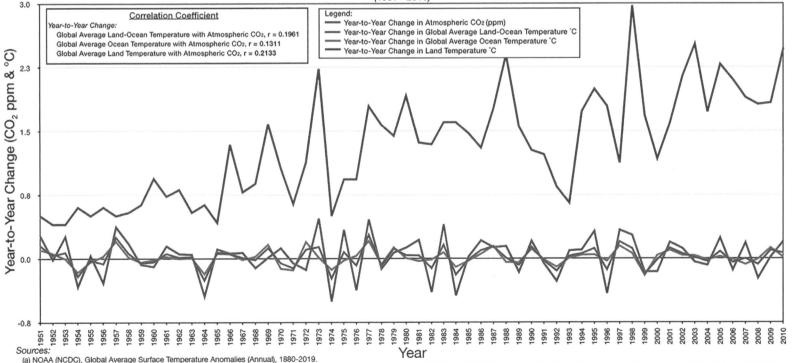

Figure 4-14
Global Average Atmospheric CO_2 (ppm) and Global Average (°C) for Combined Land-Ocean, Ocean, and Land Surfaces
(1951 - 2010)

Sources:
(a) NOAA (NCDC), Global Average Surface Temperature Anomalies (Annual), 1880-2019.
(b) ESS-DIVE (archives for Carbon Dioxide Information Analysis Center, CDIAC), Annual anthropogenic CO_2 emissions, (2014), Global CO_2 Emissions from Fossil-Fuel Burning, Cement Manufacture, and Gas Flaring: 1751-2014.

Listed on Figure 4-14 from 1951 to 2010 are the correlation coefficients for all three global average surface temperature sets. The computed correlations range between 0.1311 and 0.2133, once again confirming that changes in atmospheric CO_2 and temperature (climate) are strongly uncorrelated. As a consequence, *no causation is possible and changing atmospheric CO_2 cannot be the cause of climate change!*

Regardless of the time period chosen (139, 130, 75, 64, or 60 years), the recorded evidence relentlessly testifies the prosecution's climate change theory is invalid.

In short, *looking out the window* at the testimony of the records for global CO_2 change and global temperature change does not support the prosecution's theory.

On the basis of this evidence alone, it is extremely misleading for the prosecution to claim rising (changing) atmospheric CO_2 causes global climate warming (temperature change). The claimed relationship is impossible because the two quantities are consistently and strongly uncorrelated.

Given the many billions in funds spent on climate change research over the past three decades, shouldn't the jury reasonably expect that *somebody* from the prosecution might have mustered the courage to *look out the window* at the real-world evidence to verify that some evidence supportive of its theory-based claims *actually* exists?

Yet there is no evidence the prosecution *has ever examined these records* to confirm its theory. If it has, it certainly hasn't shared its findings with the jury. Is it deliberately withholding exculpatory evidence?

Examining 1951 to 2010, Figure 4-15 is a double-bar chart style similar to Figures 4-4 and 4-7. As with prior similar figures, this chart is based on the history of annual global average atmospheric CO_2 change together with the history of annual global average temperature change.

Figure 4-15
Does *Direction* of Annual CO_2 Change Match *Direction* of Annual Temperature Change?
(1951 - 2010)

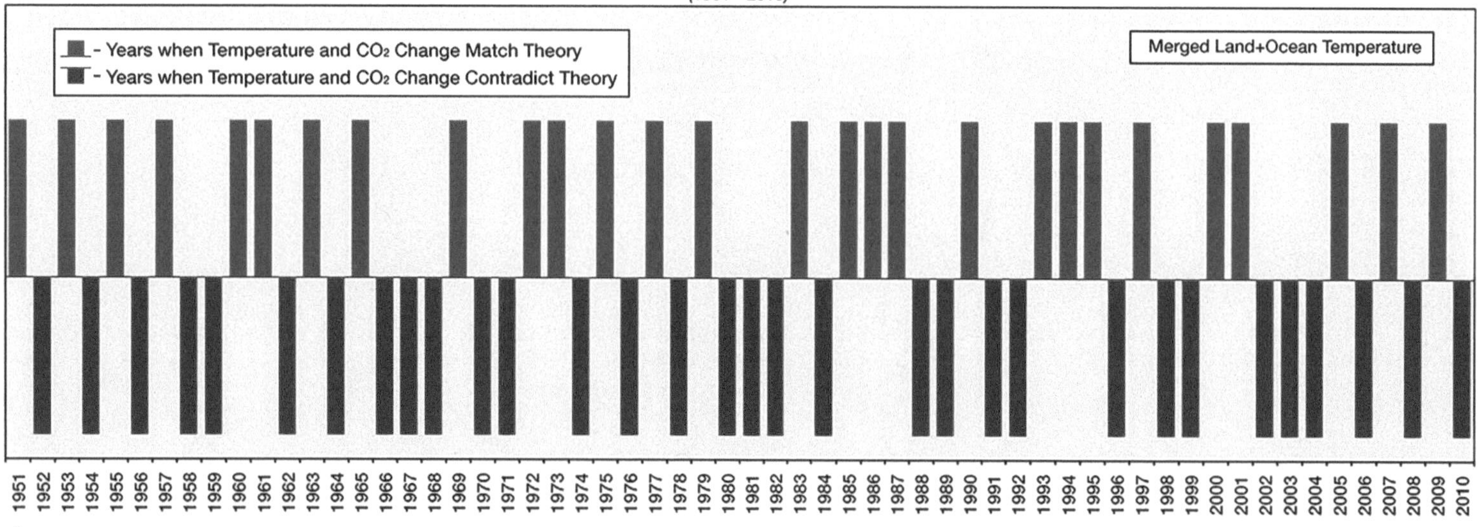

Sources:
(a) NOAA (NCDC), Global Average Surface Temperature Anomalies (Annual), 1880-2019.
(b) ESS-DIVE (archives for Carbon Dioxide Information Analysis Center, CDIAC), Annual anthropogenic CO_2 emissions, (2014), Global CO_2 Emissions from Fossil-Fuel Burning, Cement Manufacture, and Gas Flaring: 1751-2014.

When bars (green) are above the line, the annual changes are *consistent with* prosecution's theory; where the bars (red) are below the line, the annual changes are *inconsistent with* theory.

If the prosecution's theory were valid, then in accordance with that theory a large preponderance of bars should be *above* the horizontal line. Yet as the real world testifies, a *minority* of years (47%) are consistent with the prosecution's theory.

This record should particularly be theory-consistent (growing atmospheric CO_2 warms temperature; declining CO_2 cools temperature) as it covers the time range touted by the prosecution as providing strong evidence supporting its theory!

Remember, if the prosecution's radiant-heat-based greenhouse gas is such a powerful climate warming force, warming or cooling should occur virtually instantaneously with increased or decreased CO_2. Over the course of a full year of seasonal changes, the temperature should dutifully follow theory if the theory were valid. But the theory is clearly invalid for being grossly inconsistent with the record.

While 47% of the time the record is theory consistent, 53% of the time the record contradicts theory! This record is a substantial departure from theory. It is so out of line with theory that the only reasonable conclusion jurors can draw from this testimony is that *the scientific method* invalidates the prosecution's theory for its gross inconsistency with real world evidence.

Similar to what is observed over longer time periods, less than a fifty-fifty chance of agreement with theory is observed. Figure 4-15 clearly contradicts the prosecution's greenhouse gas theory of climate change. In fact, Figure 4-15 decimates the prosecution's notion that atmospheric CO_2 has a persistent strong role in any observed climate warming.

And yet, the prosecution specifically alleges to the jury that it is "extremely likely"[45] that anthropogenic CO_2 emissions were responsible for growing atmospheric CO_2 that, in turn, is alleged to be responsible for "more than half of the observed increase in global average surface temperature from 1951 to 2010…"[45] Yet real-world testimony says the prosecution's climate change theory is invalid and its theory-based claims are unsupported by real-world evidence!

Just as with Figures 4-5 and 4-8, Figure 4-16 shows the percentage of years when the prosecution's theory is consistent with the recorded observations. If the prosecution's theory were valid, then conformity of temperature should be deep in the zone of theory plausibility (> 90%). Yet once again, the 47% theory agreement with real-world evidence is found to be in the chance region, clearly indicating the prosecution's causal theory is invalid.

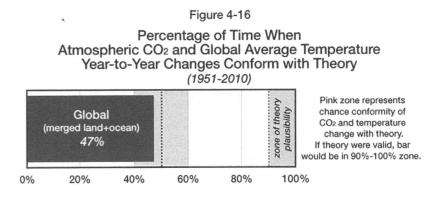

Figure 4-16

Percentage of Time When
Atmospheric CO₂ and Global Average Temperature
Year-to-Year Changes Conform with Theory
(1951-2010)

Figure 4-16 leads to the same interesting conclusions, as do prior similar charts:

- Real-world records for global averages of atmospheric carbon dioxide and near surface atmospheric temperatures are more likely to contradict than conform with climate change theory. Theory conforms with records just 47% of the 60 years between 1951 and 2010. To be valid, theory conformity should consistently occur at least 90% of the time (with a good explanation for any nonconformity during the other 10% of time).

- The evidence strongly demonstrates that when temperature change corresponds with CO_2 change it does so purely by chance.

Regardless of the prosecution's oft-repeated claims to the contrary it is abundantly clear that changes in atmospheric CO_2 *do not* produce corresponding changes in global average temperature consistent with atmospheric CO_2 being a strong climate change force.

Just how unusual are rising temperatures?

A prevalent belief nurtured by the prosecution is that late-20th-century climate is warming in response to the growth of atmospheric CO_2, and such warming will lead to an unprecedented increase in global average temperature.

Is recent climate change *really* unusual as the prosecution alleges? Or is the prosecution just fearmongering?

In Table 4-1 (page 73), the defense has already provided compelling testimony from records between 1880 and 2018 clearly showing that despite the growth of atmospheric CO_2 after 1944 being 445% higher than the rate of growth between 1880 and 1944, *climate changed at virtually the same rate in both periods*, belying the claims of the prosecution's theory!

Further theory-invalidating evidence is found in two 41-year snapshots (Figures 4-17 and 4-18) of temperature anomalies, one from the first half of the 20th century (1904–1944), the other from the second half of the century (1958–1998). Recall that anomalies are deviations from a long-term average and, as such, merely replicate the temperature change profile. Figures 4-17 and 4-18 show temperature anomalies on the same time and temperature scales. Look carefully at these charts. They each represent a 41-year progression of temperature deviations from the same long-term average.

Based on prosecution claims, jurors might be inclined to believe that observed changes in global average temperatures during the latter half of the 20th century were remarkable, unusual, and worthy of special consideration.

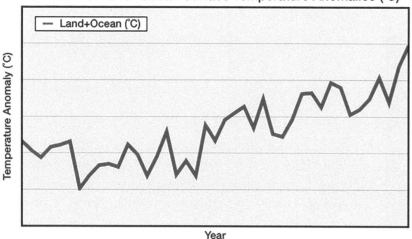

Figure 4-17
41 Years of Land+Ocean Surface Temperature Anomalies (°C)

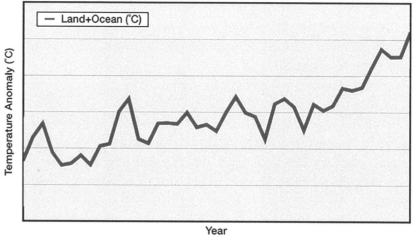

Figure 4-18
41 Years of Land+Ocean Surface Temperature Anomalies (°C)

Recall the exceptional growth of atmospheric CO_2 in the latter half of the 20th century compared with far more restrained growth in the first half of the century. Consistent with the substantial growth of atmospheric CO_2 over the latter half of the 20th century, is there really a sound basis for the prosecution's claim that global climate warming has been unusual and notably warmer than it was in the first half of the century?

Can the jury be absolutely certain which chart represents temperature from 1904 through 1944 (*before* dramatic CO_2 growth) and which chart represents temperature from 1958 through 1998 (a time during which atmospheric CO_2 grew far more rapidly to much higher atmospheric concentrations)?

If jurors cannot be absolutely certain which chart represents which time period, they must concede that the prosecution's theory-based claim is inconsistent with the observed evidence.

That there is no significant difference in temperature change while atmospheric CO_2 growth is remarkably different is highly contradictory to the prosecution's climate change theory, suggesting any claims based on that dubious theory are unsupportable.

For each 41-year period, Table 4-2 shows total temperature range (maximum temperature less minimum temperature) and net temperature growth (the net temperature change over the time period). Also shown are the differences (1958–1998 less 1904–1944) for both temperature range and net temperature growth (change).

Have jurors decided which graph likely represents which 41-year period?

Answer: Figure 4-17 is for 1958–1998 and Figure 4-18 is for 1904–1944.

Table 4-2
Two 20th Century 41-Year Periods Compared

	Early 20th Century 1904-1944	Late 20th Century 1958-1998	Difference (1958-1998) - (1904-1944)
41-Year Temperature Range (°C)	0.74	0.79	0.05
41-Year Temperature Growth (°C)	0.71	0.52	-0.19

In the absence of any other information, is it really possible to be *certain* which is which from just viewing these two graphs?

Not only is the 41 years of temperature change cooler by -0.34°F (-0.19°C) over the latter 41 years of the 20th century, 1958-1998, it was cooler while atmospheric CO_2 was growing at a rate 404% greater than it grew in the earlier, warmer period, from 1904 to 1944. This behavior is dramatically contradictory to the prosecution's alleged relationship between atmospheric CO_2 and temperature.

It is worth noting that 1958–1998 overlapped the peak of a solar grand maximum and ended during an unusually strong El Niño event, suggesting net warming (1998 temperature minus 1958 temperature) may be overstated as a consequence of these well-known natural warming forces.

In fact, there isn't a scintilla of real evidence that atmospheric CO_2 or any anthropogenic activity has ever had any real impact on global climate change over any time period for which reasonably good records exist.

This testimony clearly shows global temperature change is independent of changing atmospheric CO_2.

Any claim that "more than half of the observed increase in global average surface temperature…was caused by the anthropogenic increase in GHG concentrations"[45] is not only unsupportable with observed records, it is in fact dramatically contradicted by observed records.

Yet the prosecution has the audacity to claim a 99.9999%[67] certainty that atmospheric CO_2 growth from fossil fuel use is the primary cause of recent observed climate warming! The evidence makes a complete mockery of that dubious claim.

The 137 Years of Theory Failure (1882–2018)

Does 137 years of year-to-year atmospheric CO_2 change and year-to-year temperature change from (1882 through 2018) support climate change theory?

Figure 4-19 (next page) records those years when year-to-year atmospheric CO_2 change and year-to-year temperature change are theory compliant (changing atmospheric CO_2 relates to a corresponding temperature change).

Throughout the 137 years of atmospheric CO_2 and GAST change, while 54% of years contradicted theory, just 46% of years match theory. This theory-contradictory outcome is similar to the chance relationships between annual atmospheric CO_2 change and annual temperature change seen with similar Figures 4-4, 4-7, and 4-15.

Clearly, no causal relationship is possible between changing atmospheric CO_2 and changing temperature (climate) when theory compliance (green bars) and theory contradiction (red bars) can be emulated by the flip of a coin!

These figures testify to there being neither correlation nor causation between year-to-year atmospheric CO_2 change and year-to-year temperature change.

Just because atmospheric CO_2 is growing during periods of climate warming doesn't mean the growing CO_2 has any causal relationship with the warming. The actual real-world record makes a mockery of the prosecution's theory.

This evidence is incontrovertible proof that the prosecution's theory is unviable, invalid, and not remotely supported by real-world observed evidence.

Which should jurors believe?

- The clear evidence before their own eyes showing climate change and atmospheric CO_2 change are unrelated?
- The prosecution's climate change *theory* whose model projections testify against their own underlying theory? (see Chapter 5, "Who are the Real Climate Deniers? More Defense Testimony")

The 108 Years of Theory Failure (1911–2018)

Thirty years is sufficient time to represent climate. To examine changing climate, jurors will need to examine a history of atmospheric CO_2 and temperature change over 30-year increments. By observing the relationship between 30 years of atmospheric CO_2 change and corresponding 30 years of global temperature change (climate change), jurors can determine whether or not their behavior supports or contradicts climate change theory. By calculating a history of the 30-year changes for each ending date, a view of "changing climate" is created since each year is associated with the prior thirty years of temperature and atmospheric CO_2 change.

Figure 4-19
During 137 Years, How Often Does the *Direction* of Annual CO₂ Change Match the *Direction* of Annual Temperature Change?
(1882 - 2018)

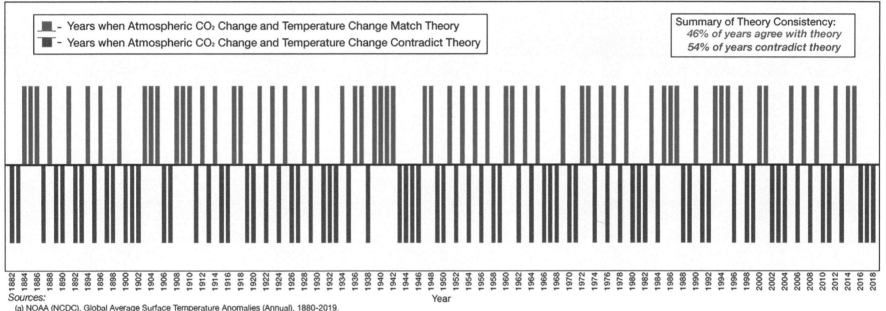

Sources:
(a) NOAA (NCDC), Global Average Surface Temperature Anomalies (Annual), 1880-2019.
(b) NOAA Earth System Research Laboratory (ESRL) carbon dioxide records, https://www.esrl.noaa.gov/gmd/ccgg/trends/data_html

From 1910 to 2018, Figure 4-20 graphs the yearly calculated thirty years of temperature change (climate) with the corresponding calculated thirty years of atmospheric CO_2 growth. For comparison, annual recorded atmospheric CO_2 (red dashed line) is also plotted.

Figure 4-20 clearly shows the strong climate warming of the 1930s followed by the four-decade climate cooling from 1940 into the 1970s.

Curiously, at a time when climate catastrophe is being heralded, the nearly 40 years of climate change history (1980-2018) shows climate is virtually unchanged in 2018 from what it was in 1980 (shown as a period of temperature stasis, yellow line). Also apparent is the lack of any consistent relationship between climate change and either yearly atmospheric CO_2 change or yearly change in 30-year atmospheric CO_2 change.

During the multi-decade cooling from the 1940s to the late 1970s climate (30-year temperature change) cooled -1.0°C (-1.8°F). Note that climate cooling during that period was enhanced by the prior sharp temperature rise during the decade of the 1930s that immediately preceded the multi-decade cooling.

It is important for jurors to distinguish between graphing year-to-year global average surface temperature change and graphing the year-to-year change of thirty-year temperature change (climate change).

Figure 4-20
108 Years of Climate (*30-year temperature change*) and Atmospheric CO₂ Growth
(1910 - 2018)

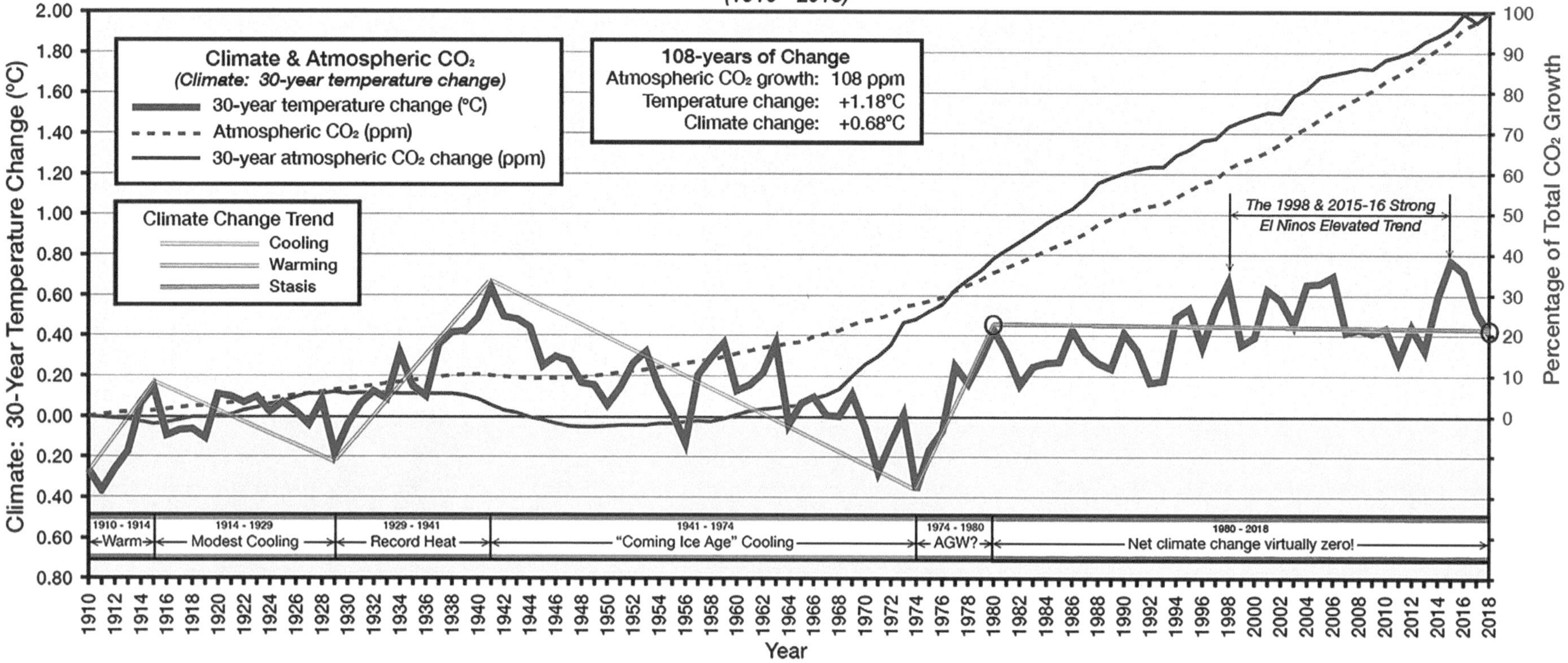

Underlying Data Sources: National Climate Data Center (NOAA/NCDC) Global Average Surface Temperatures (GAST); Mauna Loa Observatory & CDIAC/ESRL (CO₂)

This history of "changing climate" can be examined to view how well it agrees with the prosecution's climate change theory. Figure 4-20 shows that post-1980 climate change is consistent with the temperatures recorded by the 114 new carefully-sited US Climate Reference Network (USCRN) observing stations that not only show no warming since they began reporting in 2005, they show *a net cooling* from 2005 through 2019. More detailed information about the USCRN records is described in both Chapter 5 and Defense Exhibit B.

Based on annual increments of 30 years of atmospheric CO_2 change compared with the annual increments of climate change (30-year temperature change) from 1911 to 2018, Figure 4-21 provides jurors powerful evidence contradicting greenhouse gas climate change theory by showing the yearly relationship between *30 years of atmospheric CO₂ change* and *30 years of temperature change* (climate) across the 108 years from 1911 through 2018.

For each year of Figure 4-21, agreement with theory occurs whenever the direction of 30-years of atmospheric CO_2 change ending on that year is the same as the direction of 30-years of temperature change ending on that year.

Figure 4-21
During 137 Years,
How Often Does the *Direction* of 30-year Atmospheric CO_2 Change Match the *Direction* of 30-year Temperature (Climate) Change?
(1882 - 2018)

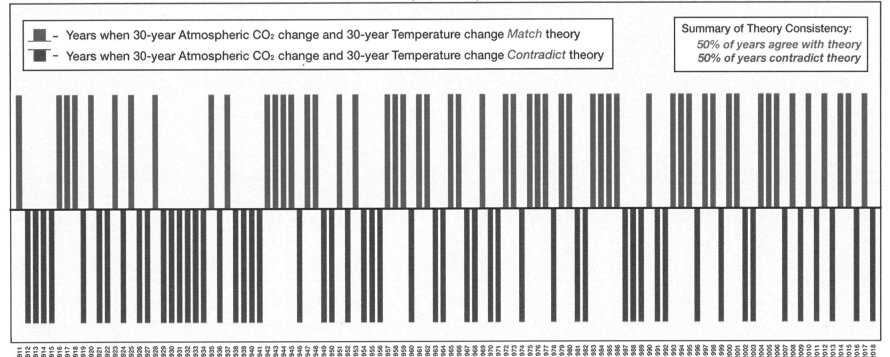

■ – Years when 30-year Atmospheric CO_2 change and 30-year Temperature change *Match* theory

■ – Years when 30-year Atmospheric CO_2 change and 30-year Temperature change *Contradict* theory

Summary of Theory Consistency:
50% of years agree with theory
50% of years contradict theory

Year

Sources:
(a) NOAA (NCDC), Global Average Surface Temperature Anomalies (Annual), 1880-2019.
(b) NOAA Earth System Research Laboratory (ESRL) carbon dioxide records, https://www.esrl.noaa.gov/gmd/ccgg/trends/data_html

For example, for each year of this chart, if *both* 30 years of CO_2 and temperature changes increase or decrease (i.e., change in the same direction), then a green bar showing theory agreement is charted. On the other hand, if 30 years of CO_2 change increases while 30 years of temperature change decreases (or 30-year CO_2 change decreases while 30-year temperature change increases), then a red bar showing theory contradiction is charted.

Just as jurors observed with annual change, increments of 30 years of CO_2 and climate change have no better agreement with theory than does year-to-year CO_2 and temperature change.

Not surprisingly, the relationship of changing climate to changing atmospheric CO_2 is observed to be one of pure chance equivalent to a coin flip. Once again, prosecution's theory is soundly contradicted.

Characteristics of the Contemporary Records (1880–2010)

Spanning 30, 32, 33, and 35 years of the 130 years between 1881 and 2011, in Chapter 1, jurors examined four different climate periods to observe the relationship between the changing atmospheric concentration of CO_2 and global climate change (as represented by global average merged land and ocean surface temperatures).

Figure 4–22 displays the full 130-year record of those same temperatures and atmospheric CO_2 showing the four climate periods shaded according to whether climate cooled or warmed. Note that homogenization of the temperature record (examined in Chapter 5) served to cool the record heat of the 1930s, moderate the strong cooling between 1944 and 1976, and exaggerate the warming after 1976. (For more about homogenization of temperature records, see Chapter 5 "Is *Evidence of Man-Made Climate Change* Really *Man-Made Evidence of Climate Change?*")

Table 4-3
130 Years of Compelling Evidence: Atmospheric CO₂ Change *Is Not* A Climate Change Force
(1881 through 2011)

Period	Years	ΔCO₂ (ppm)	ΔCO₂ (ppm/yr)	ΔT (°C)	ΔT (°C/yr)
1881 - 1911	30	+9.2	+0.31	-0.37	-0.012
1911 - 1944	33	+9.6	+0.29	+0.73	+0.022
1944 - 1976	32	+21.8	+0.68	-0.38	-0.012
1976 - 2011	35	+59.6	+1.70	+0.66	+0.019
1881 - 1976	95	+40.6	+0.43	-0.02	-0.000
1881 - 2011	130	+100.2	+0.77	+0.64	+0.005

Source: Figure 4-22, Combined Land+Ocean Temperature & Atmospheric CO₂

Each of the 30- to 35-year periods displayed an entirely different relationship between changing atmospheric CO_2 and changing global average temperatures. Table 4–3 shows that while atmospheric CO_2 persistently grew, climate cooled, then warmed, then cooled, then warmed again. The 1881–1976 net change was a very slight *cooling*!

Based on the 130-year net change, 100 years of climate warming was just 0.49°C (0.88°F), typical for natural post-LIA climate change entirely unrelated to atmospheric CO_2.

The prosecution ignores the evidence between 1881 and 1976 to mislead the jury with the post-1976 record. Yet science doesn't change over time. If the prosecution's theory is valid for 1976 through 2011, then it should have been valid in each of the three preceding periods of climate change from 1881 through 1976. But it isn't.

Holocene Interglacial versus Climate Change Theory

Figure 4–23 is derived from *An Overview to Get Things into Perspective*[68] (by Ole Humlum, professor emeritus, Institute for Geosciences, University of Oslo).

This chart is created from the Greenland GISP2 (Greenland Ice Sheet Project 2) ice core records that graphically show the 10,700-year history of temperature and atmospheric CO_2 during the current (Holocene) interglacial.

Given the 10,700-year time period involved, these curves represent changing *climate* as represented by temperature.

Based on GISP2 ice core analyses establishing the relationship between well-mixed atmospheric CO_2 and temperature, these curves end just

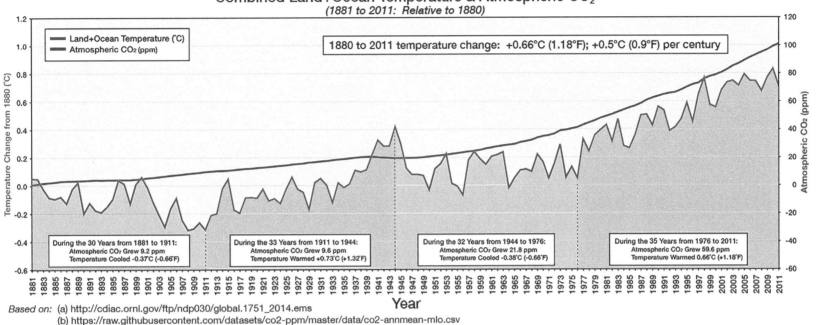

Figure 4-22
Combined Land+Ocean Temperature & Atmospheric CO₂
(1881 to 2011: Relative to 1880)

Based on: (a) http://cdiac.ornl.gov/ftp/ndp030/global.1751_2014.ems
(b) https://raw.githubusercontent.com/datasets/co2-ppm/master/data/co2-annmean-mlo.csv

prior to the global records of the late 19th century through 2018, previously examined throughout most of this chapter.

The small red block to the left of the zero on the time scale roughly indicates the past 150 years (labeled "Modern warm period") as Earth warmed following the end of the Little Ice Age (circa 1380-1850).

This graphic gives the jury a clear picture of powerful additional testimony for the defense.

The prosecution alleges greenhouse gas climate change is (1) global in nature and (2) its impacts should be felt more severely and sooner in polar regions (this is known as *polar amplification* and is examined in greater detail in "Polar Amplification: Another *Must Be* of Climate Change Theory *That Isn't!*" in Chapter 5).

These tenets of greenhouse gas climate change theory give greater weight to the testimony of Greenland's ice core record. Why?

Because there are only two alternatives for interpreting this information, and *both views* are fatal to the prosecution's climate change theory:

Alternative 1: The theory is valid, but the impact predicted is not occurring because atmospheric CO₂ growth is insufficient to produce the theory-required effect.

Alternative 2: The theory is invalid, which explains why the predicted effect of the theory is nonexistent.

The second alternative is the only reasonable choice based on the weight of the accumulated defense testimony coupled with the evidence in Figure 4-23 that clearly contradicts the prosecution's theory-based relationship between temperatures (blue curve above) and atmospheric CO_2 (red curve below).

Over the thousands of years of the Holocene interglacial, as CO_2 falls and then rises, temperature rises and then falls!

The observed relationship is just as contradictory to the greenhouse gas climate change theory as is the evidence on the 550-million-year scale (Figure 3-5, page 46) shown for the period between 470 mya and 430 mya when global climate cooled 10°C (18°F) while atmospheric CO_2 grew dramatically, and then as climate warmed 10°C (18°F), atmospheric CO_2 precipitously declined by nearly four times the current level of CO_2 in the atmosphere!

Once again, jurors have the clear testimony of real-world evidence that unequivocally contradicts the prosecution's climate change theory.

Ironically, if the atmospheric CO_2 chart (red curve, bottom of Figure 4-23) were inverted, it would be a better (though still unconvincing) representation of the prosecution's flawed greenhouse gas climate change theory.

This evidence alone is sufficient for *the scientific method* to reject greenhouse gas theory for yet again failing to conform with real-world evidence.

Ironically, this evidence supports the view that atmospheric CO_2 growth leads to *climate cooling*, just the opposite of the prosecution's climate change theory.

Finally, the prosecution's claim of dramatic post-1940 melting of southeastern Greenland's continental glacial is soundly refuted by the evidence.[69]

There is incontrovertible proof that southeastern Greenland gained significant glacial mass (and thickness) during the post-1940 period from 1942 to 1988 while atmospheric CO_2 was skyrocketing!

During World War II (in 1942), about the time global average atmospheric CO_2 began its dramatic growth, eight military aircraft left the USA for Britain (*condensed quote follows*):

> ...[A] P38 Lockheed Lightning now known as *Glacier Girl* was one of six P38 fighters and two B17 bombers on their way to Britain. Bad weather led to their forced landing on the ice cap. They were located in 1988, over 80m, (or 264 feet) deep in the ice and about 2 miles away from the crash site. Recovery of *Glacier Girl* was started a couple of years after discovery allowing another couple of metres of snow and ice to add to its cover. So we have undeniable, documented, physical evidence—the aircraft has been restored and is flying—the SE area of Greenland was not melting away or losing ice cover at all, at least not before 2003/4.[69]

In fact, southeast Greenland's glacial cover had been increasing dramatically throughout the period when global atmospheric CO_2 was also increasing dramatically!

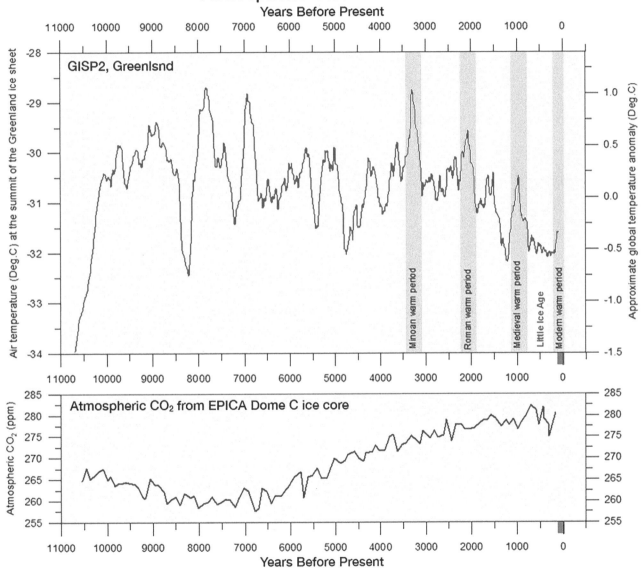

Figure 4-23

Holocene Interglacial Temperatures at Greenland Ice Sheet Summit
Atmospheric Carbon Dioxide

Source: http://www.climate4you.com/GlobalTemperatures.htm

Conclusion

This chapter's comprehensive examination of the real-world relationship between changing atmospheric carbon dioxide and changing temperature is based on both recorded evidence dating from the late 19th century and ice core analyses of Greenland's year-round continental glacier. This testimony exonerates not only atmospheric CO_2, it exonerates *every source of atmospheric CO_2 growth* while unequivocally rejecting the prosecution's greenhouse gas climate change theory that is found to be contradicted by real-world evidence and thus invalidated by virtue of its failure to satisfy requirements of *the scientific method.*

Flawed theory renders the prosecution's theory-based models similarly flawed (more about model failures in Chapter 5).

Discarding this seriously flawed theory is not only demanded by *the scientific method* protocol, it is long overdue!

Recall the prophetic words of Dr. Marcel Leroux:

> The most urgent priority for climatology…is to leave the IPCC…

Summary: *Global Average Surface Temperature and Atmospheric CO_2*

1. Persistent and compelling contradictions to the prosecution's theory are found in the testimony of records for atmospheric CO_2 and climate change. Neither annual nor 30-year atmospheric CO_2 change has any detectable causal relationship with corresponding annual or 30-year climate change to support any charges against either human activity or atmospheric CO_2 as a contributor to climate change.

2. The testimony of real-world evidence reveals the *Paris Agreement* and *Green New Deal* to be costly folly that are entirely inappropriate. Reducing CO_2 emissions will have no impact on global climate. Allegations to the contrary are unsupportable with evidence and defy the clear testimony of the long history of real-world evidence.

3. CO_2 growth is not the source of post-LIA climate change. The actual cause of CO_2 growth is unimportant because CO_2 growth is observed to have no causal relationship with climate change in any meaningful record of atmospheric CO_2 and climate. Consequently, since growing atmospheric CO_2 is not a meaningful climate warming force, fossil fuel emissions are entirely irrelevant!

4. That both CO_2 and temperature have a long-term net increase over the post-1850 span of climate history is either by chance or because they are both responding to another stimulus (possibly solar activity). There is not a scintilla of real-world evidence that atmospheric CO_2 growth is either correlated with or has any causal relationship with climate warming. Claims to the contrary are rooted in speculative theory that is invalidated by observed real-world evidence.

5. More compelling evidence contradicting climate change theory is testified to by the pronounced cooling trend between 1941 and 1976 during which atmospheric CO_2 was sharply *growing by 21 ppm* at a time when global temperatures were cooling by 0.49°F (0.27°C) and 30-year temperature change (climate change) was cooling by 1.8°F (1.0°C) raising fears of "another ice age".[70]

6. The long-term relationship between changing atmospheric CO_2 and temperature change is indistinguishable from pure chance, clearly invalidating the prosecution's theory that virtually locks atmospheric CO_2 growth to climate warming and atmospheric CO_2 decline to climate cooling.

7. For decades, the jury has repeatedly been told that growing atmospheric CO_2 is a strong climate change force responsible for global warming. The jury has also been told that atmospheric CO_2 growth is caused by modern civilization's use of fossil fuels. The testimony of real-world records for atmospheric CO_2 and temperature unequivocally rebut both these prosecution allegations, showing them to be completely lacking support in real-world evidence. The accusation that fossil fuel use is responsible for atmospheric CO_2 growth is entirely irrelevant since *total* atmospheric CO_2 change at concentrations typical of the past half billion years has never been observed to be a detectable climate change force. Therefore, the source of atmospheric CO_2 growth is entirely immaterial and irrelevant.

8. Any climate change theory unsupported by observation (the recorded real-world evidence revealed by *looking out the window*) is necessarily an invalid theory that, in accordance with *the scientific method, must be rejected* for failure to agree with real-world observations.

9. Comparing two 41-year spans of global average temperature change, one from the early 20th century and one from the latter decades of the 20th century, jurors have seen their similarity is such that neither could accurately be characterized as being either "unprecedented" or unusual and neither portends any "catastrophic" warming. This evidence suggests climate behaved similarly in two distinctly different periods of CO_2 growth, one during a very low rate of atmospheric CO_2 growth and the other during a persistent dramatic increase in the rate of atmospheric CO_2 growth. Contrary to theory, temperature change during the 41 years of the latter part of the 20th century was actually *27% less than* the temperature change during the 41 years in the early 20th century when much lower atmospheric CO_2 growth

prevailed. These observations render the prosecution's climate change theory invalid.

10. Greenland (GISP2) ice core analyses reveal that during the 10,700 years of the current (Holocene) interglacial, as atmospheric CO_2 declined and then grew, climate warmed and then cooled—*exactly contradictory* to the prosecution's theorized climate change relationship that emphasizes polar amplification (greater and earlier polar effects). This is further compelling evidence that the prosecution's greenhouse gas climate change theory is hopelessly invalidated by simply *looking out the window* at the evidence.

11. Those who are still unconvinced are free to examine official US government global temperature and CO_2 records.[1, 3, 6, 7, 66] Aided by the observations and conclusions based on that evidence, jurors should have no difficulty concluding the evidence clearly invalidates the prosecution's climate change theory, making the source of atmospheric CO_2 growth entirely irrelevant.

12. At the beginning of this chapter, the prosecution claimed, *"Recent climate warming is principally caused by atmospheric CO_2 growth."* An examination of real-world evidence not only fails to confirm that claim, it unequivocally rejects it.

13. Based on the testimony of US government records for both global atmospheric CO_2 and global average surface temperatures, the conclusion reached in *Role of Greenhouse Gases in Climate Change* (2017) by Hertzberg and Schreuder[50] is fully supported:

> Atmospheric CO_2 *is not a significant climate change force* and anthropogenic emissions of CO_2 (including fossil fuel CO_2) are a barely detectable fraction of atmospheric CO_2 growth. Consequently, *human activity has no discernible*

impact on global climate change that is well within natural climate variability.

14. The prosecution's greenhouse gas climate change theory is revealed to be entirely unsupported by any consistent real-world evidence. As such, it fails a key provision of *the scientific method* and is invalidated. Any study or action based on the prosecution's invalid theory lacks a legitimate foundation and should not be taken seriously. Adding insult to injury, the prosecution's own climate simulation models invalidate the prosecution's theory upon which they are based (see Chapter 5, *"Who are the Real Climate Deniers? More Defense Testimony: When Computer Models Get It Wrong, Question Their Basis"*).

15. The evidence presented in this chapter together with compelling additional evidence presented in Chapter 5 will leave the jury with little recourse but to declare atmospheric CO_2 innocent of any capacity to alter global climate to any discernible degree and on any consistent basis. There simply is no evidence in any records for any period of time that supports the prosecution's theory upon which today's highly politicized climate change narrative is based.

16. Given the complete lack of any supporting evidence for the climate change theory promoted by the UN-IPCC (the prosecution), the US government was absolutely correct in both its decisions to (1) enhance US production and use of fossil fuel energy and (2) quickly withdraw the US from participation in the *Paris Climate Agreement*, an agreement lacking a solid scientific foundation and whose objectives have no chance of discernibly impacting global climate.

17. For being deceived for decades by a specious theory that contradicts real-world evidence, governments of nations around the globe should re-examine their participation in the *Paris Climate Agreement*, their support for the UN-IPCC, their

own carbon restriction regulations, and climate change material based on a *deeply flawed theory* being taught in their schools and universities.

18. Based on the compelling unequivocal testimony of real-world evidence found in atmospheric CO_2 and climate histories over every time period examined (geologic, ice core, contemporary), there is compelling evidence that, regardless of the source of any change, changing atmospheric CO_2 has no detectable impact on global climate change.

19. Because the prosecution's climate change theory is invalid, the climate change narrative supported and promoted by agents of the prosecution for nearly four decades is also invalid and is not a legitimate basis for any national or international policy or regulations designed to limit fossil fuel production or usage.

20. There is no evidence in nature that at current and future concentrations, changing atmospheric CO_2 has any capacity to alter global climate.

21. For those jurors who are not yet convinced, take another look at Table 4-1 (page 73) and Figures 4-9 and 4-10 (pages 72 and 74, respectively). Try to find a theory-consistent explanation for the clear testimony of this evidence. These straightforward observations offer compelling testimony that the prosecution's climate change theory lacks scientific credibility.

CHAPTER 5

Who are the Real Climate Deniers?
More Defense Testimony

When Inconvenient Records Contradict Greenhouse Gas Theory

In Chapter 4 the defense carefully examined extensive records of changing atmospheric carbon dioxide and temperature in search of any real-world evidence that supports a cause-effect relationship consistent with the prosecution's climate change theory. Jurors discovered the observed evidence shows no such cause-effect relationship exists because the two are strongly uncorrelated. In fact, the actual observed relationship between changing atmospheric carbon dioxide and temperature is one of pure chance (the flip of a fair coin).

While the defense believes it has unequivocally proven the prosecution's theory lacks validity and, therefore, the case against atmospheric CO_2 is specious and without merit, nevertheless, jurors should be aware of additional compelling evidence supporting the defense case. Chapter 5 examines this additional evidence that challenges the pros-

ecution's theory of a CO_2 climate crime, including testimony that observed evidence does not confirm the existence of several science-required consequences of the prosecution's climate change theory.

The Inconvenient Pause: What Warming Trend?

With the exception of a late 20th/early-21st century solar grand maximum, and two particularly strong El Niño Southern Oscillation (ENSO) events (1998 and 2016), no *significant* warming trend is evident in either radiosonde measurements (weather balloon) or University of Alabama-Huntsville (UAH) satellite measurements[71-72] (1979 through 2018) of Earth's global lower atmosphere (Figure 5-1). A modest ~0.3°C jump across the 1998 ENSO event concurrent with a strong solar maximum is evident. Observed warming is magnified by these strong warming impulses. Note the relative insensitivity to the 1998 ENSO warming spike in the weather balloon record (shown in green) during this time. The satellite record is free of the notorious characteristic errors that plague legacy surface station records.

When ENSO warming spikes are not displayed (the 1998 warming spike is removed in Figure 5-2, page 96), *there is no evidence of any significant warming trend* during the period 1979 to 1998. If the 2015–16 ENSO spike were removed from 1999 to 2019 (Figure 5-3, page 96), no significant warming trend would be evident from the early 2000s up to 2019. After the 2016 ENSO spike, the UAH record cooled through 2018. This same cooling is seen in both the NCDC global and USCRN continental US land surface temperature records for 2016–2019.

The lack of a warming trend is supported by radiosonde records (in green, Figure 5-1, next page).[73] While both these temperature measuring methods are *more accurate and reliable* than the ground station readings used in Chapter 4, satellite and radiosonde readings are not available for the much longer timeframes of ground station records.

These graphs (Figures 5-1 through 5-3, next several pages) are modified presentations of the UAH graphs created by Dr. Roy Spencer and Dr. John R. Christy, whose graphs are regularly updated at the DRRoySpencer.com website, an excellent site for tracking this information. Figures 5-1 and 5-3 are updated through 2018 and clearly show the impact of both the 1998 and 2015–16 strong El Niño (ENSO) events.

Figure 5-1 adds the corresponding atmospheric CO_2 history (1979 to 2019) to Spencer's graph. Note that these temperature and atmospheric CO_2 records are inconsistent with the prosecution's climate change theory. This is further confirmation of the dubious nature of the prosecution's climate change theory.

The UAH satellite record shown in Figures 5-1 to 5-3 reflect the updated version 6 of the UAH processing update that is perhaps the very best and most rigorous satellite atmospheric temperature record available through 2018.[74]

Thanks to the rigorous standards maintained to assure data accuracy, weather balloon radiosonde records are highly reliable. It is worth noting that radiosonde (weather balloons) has been in use since the 1930s, reliably reporting atmospheric conditions that help improve the accuracy of weather forecasts.[75] Weather forecasters have relied upon the massive data collection and accuracy of radiosonde weather balloons for almost 90 years.

Figure 5-1 shows that between the beginning of the satellite record in 1979 and 2019, the most notable warming spikes coincide with the strong El Niño years in 1998 and again in 2015–16. Bear in mind that these are satellite-based temperature records, not global average surface temperature records (examined in Chapter 4).

Between 1979 and 1997 (before the 1998 El Niño), Figure 5-2 shows a very slight temperature warming trend (satellite record varies roughly +/-0.3°C around an average of about -0.15°C; radiosonde record varies by +/-1.0°C around an average of roughly -0.08°C).

Jurors should note that the measured temperature oscillates over two- to four-year periods with peaks and valleys showing a very slight upward trend. No significant temperature increase is observed, yet atmospheric carbon dioxide *steadily increases* throughout the 1979 to 2019 period. This relationship is highly inconsistent with the prosecution's climate change theory.

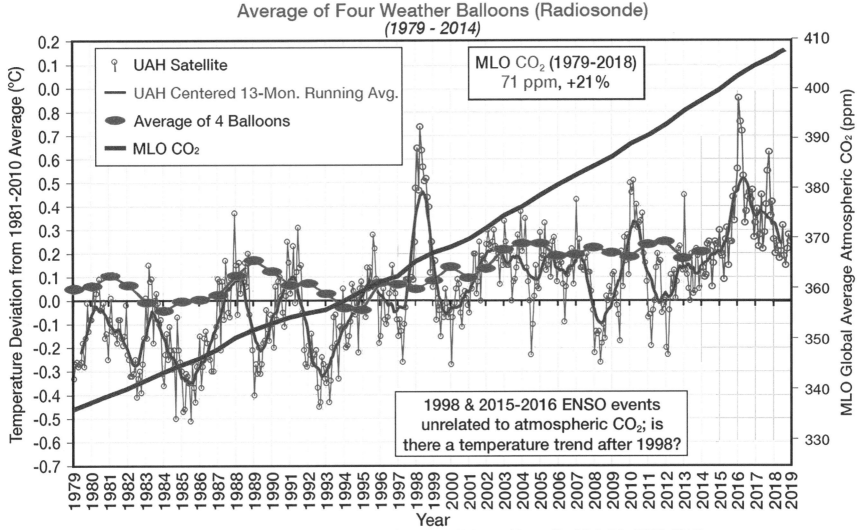

Figure 5-1

UAH (Ver 6) Satellite-Based Temperature of the Global Lower Atmosphere
(1979 - 2019)
Average of Four Weather Balloons (Radiosonde)
(1979 - 2014)

Sources: DRRoySpencer.com, Dr. Roy Spencer, Dr. John R. Christy, U. of Alabama, Huntsville; MLO CO₂ (1979-2018)

Figure 5-2
UAH (Ver 6) Satellite-Based Temperature
of the Global Lower Atmosphere
(1979-1998)

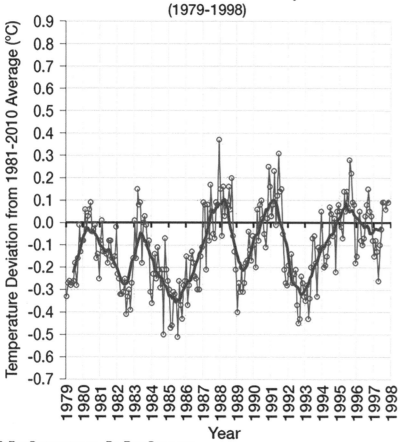

DrRoySpencer.com, Dr. Roy Spencer

Figure 5-3
UAH (Ver 6) Satellite-Based Temperature
of the Global Lower Atmosphere
(1999-2018)

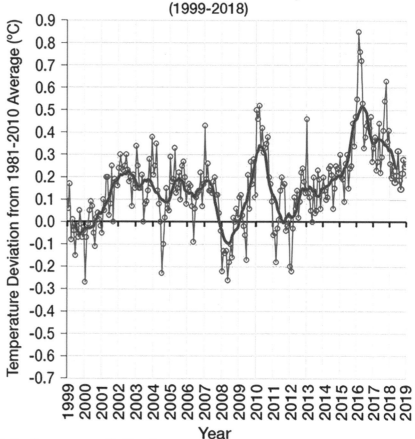

DrRoySpencer.com, Dr. Roy Spencer

Figure 5-3 shows virtually no evidence of any meaningful temperature trend during the fourteen years of the "warming pause" from 2002 to 2015 inclusive. This evidence is consistent with USCRN's new temperature-monitoring stations across the contiguous 48 US states that shows a slight cooling from 2005 through 2019 (see Figure 5-12, page 115). As this warming pause unfolded the prosecution issued fren-

zied reports warning of *human-caused climate change* when, in fact, there was absolutely no evidence of any non-ENSO-related temperature warming trend. After 2015, the strong 2015–16 El Niño spiked temperatures, but from 2016 to 2019, temperatures cooled back to their pre-ENSO level, extending the "warming pause" to nearly two decades.

A 1998 study[76] published by the American Meteorological Society (AMS) examined worldwide radiosonde measurements and concluded:

> The research work presented in this paper indicates that climate trends can currently be estimated reliably with a subset of the upper-air data.

Radiosonde temperature records (weather balloon, in green, Figure 5-1) during the 36 years from 1979 through 2014 show remarkably tame 0.15°C (0.27°F) temperature warming, at a rate equivalent to just under 0.42°C (0.75°F) warming over 100 years, somewhat *more modest* than typical post-LIA climate warming.

This suggests that if not for the strong 1998 and 2015–16 El Niño (ENSO) temperature spikes, the 41-year temperature increase in the UAH record (1979 through 2018) of ~0.55°C (0.99°F) might very likely have been *less than* observed.

Perhaps even more important, the 1998 ENSO temperature spike indicates the influence of atmospheric CO_2 on temperature change is sufficiently nonexistent that any evidence of it is easily overwhelmed by an ENSO cycle.

UAH satellite records confirm the same negligible trends before and after the temperature shift across the 1998 ENSO event. The 1998 ENSO event appears to have had a noticeable impact with lingering effects on the baseline global average temperature. Figure 5-1 shows the 41-year (1979-2019) maximum range of the 13-month running average of temperature is +0.86°C (+1.55°F) from -0.35°C to +0.51°C. Absent the 2016 strong ENSO, the maximum range of the 13-month running average is 20% lower (+0.69°C).

During the 41 years (1979–2019), the UAH satellite record (Figure 5-1) shows eight distinct declines of -0.25°C or more in the 13-month running average of temperature that produced -0.34°C (-0.61°F) of cooling. All while atmospheric CO_2 continued to grow at about the same rate (averaging 1.9 ppm/year). Did the prosecution's climate change theory take a vacation during all those cooling years?

Since the prosecution's climate change theory is radiation-based, temperature change should be virtually instantaneous with no meaningful time lag in the "back-radiation" process often described by the prosecution:

> The amount of heat radiated from the atmosphere to the surface (sometimes called "back-radiation") is equivalent to 100 percent of the incoming solar energy. The Earth's surface responds to the "extra" (on top of direct solar heating) energy by raising its temperature.[77]

Aside from this bizarre view having Earth's attempt to cool added to incoming solar radiation, the radiation and reradiation that creates the claimed "back-radiation" warming effect occurs with the speed of light. Since this is a continuous process, if this view were real, shouldn't changing CO_2 produce corresponding temperature change?

How is it possible that the record shown in Figure 5-1 can contain so many *years* of cooling while atmospheric CO_2 is consistently growing? Is the jury expected to believe that during cooling periods unspecified external forces are stronger than the prosecution's exalted greenhouse gas climate change theory, yet during warming periods theory prevails?

How can the jury be expected to believe growing CO_2 is such a strong warming force if it can be overwhelmed so frequently?

Note the virtually steady growth of atmospheric CO_2 between 1979 and 2019 (orange MLO line, Figure 5-1). Oddly, the MLO curve smoothly transitions the two strong ENSOs without a blip, right through periods of cooling alternating with warming. A pronounced uptick in CO_2 growth might have been expected from rapidly warming waters in the large El Niño-affected areas of the tropical Pacific Ocean. But no uptick is evident.

Perhaps the real focus should be on just how much of the actual temperature change shown in the satellite record might be a result of lingering effects from the ENSO cycles. This is particularly evident when the MLO CO_2 record is compared with the temperature profiles from records for both the UAH satellite and radiosonde measurements.

How is it that atmospheric CO_2 and burning fossil fuel for energy can be the cause of climate warming when both satellite and radiosonde records *independently* show no relationship between atmospheric CO_2 growth and climate warming over the entire 41-year (1979 to 2019) timeframe? Also note from 2016 through the end of 2018 the sharp three-year decline of temperature in the satellite record. That universally-recorded decline came while atmospheric CO_2 persistently grew, yet another clear contradiction to the prosecution's dubious climate change theory.

By denying the warming pause ever existed, the prosecution loses all credibility in the face of both satellite and radiosonde records (Figure 5-1) that clearly confirm the pause and show no significant deviation from the normal long-term post-LIA trend of modest climate warming.[71-73]

So who are the *real* deniers?

The prosecution, who misleads the jury while denying the evidence?

Or the defense, who consistently exposes real-world evidence-based contradictions to the prosecution's theory?

Summary: *The Inconvenient Pause—What Warming Trend?*

1. The satellite-based lower atmosphere temperature evidence shows a normal post-LIA slight upward temperature trend between 1979 and 1998.
2. The strong 1998 El Niño (unrelated to atmospheric CO_2) coincides with a sudden baseline jump of roughly 0.3°C in satellite temperature measurements.
3. The post-2002 (2002–2019) satellite evidence shows no meaningful temperature trend. A 2005–2014 period known as the "pause" is clearly evident in the radiosonde record, corroborating the satellite evidence.
4. The net satellite temperature change from 1979 to 2019 is roughly +0.53°C (+0.95°F). If the lingering 1998 ENSO effects include a baseline temperature shift of +0.3°C (as Figure 5-1 suggests), then the net non-ENSO-related post-LIA temperature growth for the 41 years from 1979 to 2019 would be 0.23°C or about 0.0058°C per year, representing a post-LIA 100-year temperature warming of 0.58°C (+1.04°F). This compares well with the land-based 0.0066°C per year.
5. Radiosonde (weather balloon) records confirm satellite readings.
6. There is no evidence of any unusual non-ENSO-related post-LIA temperature warming to justify the prosecution's campaign to demonize either CO_2 or any portion of atmospheric CO_2 growth attributable to fossil fuel use.

The Missing Hot Spot—an Inconvenient Tipoff That Climate Change Theory Is Flawed!

After crafting a climate change theory that *assumes* fossil fuel combustion is responsible for the atmospheric CO_2 growth that is *claimed* to be the primary driver of climate warming, the prosecution set out to use climate simulation models to provide evidence based on model projections of how much climate would warm, where it would warm, and for how long it would warm.

The prosecution's theory-based climate simulation models are unassailable gospel to those who view the greenhouse gas climate change theory as dogma. Models have been routinely tweaked and rerun and tweaked some more and rerun again to project the theory's view of Earth's future climate should humanity continue harnessing abundant inexpensive energy from fossil fuels.

Now, just suppose the prosecution's claims actually had merit and greenhouse gases were, in fact, actually raising global temperatures as alleged. That situation would create yet another inconvenient problem for the prosecution.

Greenhouse gas climate change theory projects *with absolute certainty* several bedrock consequences of climate warming caused by atmospheric CO_2 growth.[79] One of those *certain requirements* is a strong telltale "hot spot" of greenhouse gas climate warming that *must exist* in the tropical mid-troposphere (at altitude 8–12 km in the tropical latitudes between 30°N and 30°S).

Theory-based models dutifully predict this distinctive telltale greenhouse gas global warming hotspot shown as the red central area of Figure 5-4.[80–82] Figure 5-5 shows the models' warming prediction from all climate change forcings including greenhouse gases that clearly reveals the *greenhouse gas hot spot* (central red area) *dominates* all climate warming forces.

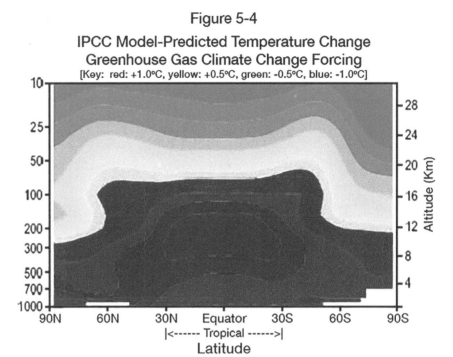

Figure 5-4
IPCC Model-Predicted Temperature Change
Greenhouse Gas Climate Change Forcing
[Key: red: +1.0°C, yellow: +0.5°C, green: -0.5°C, blue: -1.0°C]

Yet when Earth's actual atmospheric temperature is measured by weather balloons (radiosonde) to confirm the existence and extent of this hot spot, it becomes quite evident that something is amiss. The disparity between theory-based model predictions (Figures 5-4 and 5-5) and the actual measured temperature (radiosonde, Figure 5-6, next page) is a striking real-world rejection of the prosecution's climate change theory.

The real-world radiosonde and satellite measurements do not detect *even a hint* of the prosecution's theory-required greenhouse gas climate-warming hot spot. This is a clear indicator that something is very wrong with greenhouse gas climate change theory.

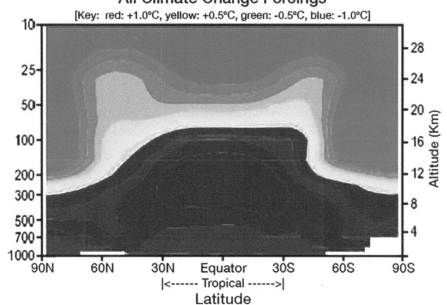

Figure 5-5

**IPCC Model-Predicted Temperature Change
All Climate Change Forcings**

[Key: red: +1.0°C, yellow: +0.5°C, green: -0.5°C, blue: -1.0°C]

Altitude (Km)

|<------ Tropical ------>|

Latitude

Figure 5-6

Actual Radiosonde Measured Temperature Change

[Key: red: +1.0°C, yellow: +0.5°C, green: -0.5°C, blue: -1.0°C]

Altitude (Km)

|<---------- Tropical ---------->|

Latitude

While theory-based models confidently predict strong greenhouse gas warming (Figures 5-4 and 5-5), real-world measurements tell an entirely different story (Figure 5-6). Not only do these real-world measurements of tropical tropospheric temperatures fail to detect the required greenhouse gas strong warming hot spot, they actually detect a modest *cool spot* (pale blue area, lower center of Figure 5-6)!

How can this be?

Because the prosecution treats its theory as dogma, it *never lets real-world observations stand in the way of its climate change narrative.* Whenever the evidence refutes the prosecution's theory, the prosecution *never reexamines* its theory; instead, it simply ignores the evidence and looks for a rationale (measurements must be faulty, it's really there, etc.).

Occasionally, the prosecution produces a study that purports to have found the illusive mandatory hot spot[83] (hallelujah!), but when carefully examined,[84] (dang!) the discovered hot spot is found to be (1) not at all as hot as the theory requires and/or (2) caused by a transitory climate dynamic not associated with greenhouse gases. So once again the prosecution simply goes back into hot spot hibernation, ignoring the important implications of the missing hot spot by implying the hot spot isn't really that important (which, of course, is contrary to its own theory).

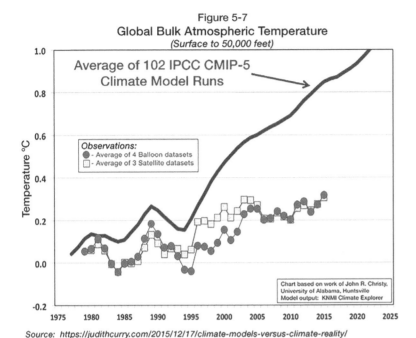

Figure 5-7
Global Bulk Atmospheric Temperature
(Surface to 50,000 feet)

Average of 102 IPCC CMIP-5 Climate Model Runs

Observations:
- Average of 4 Balloon datasets
- Average of 3 Satellite datasets

Chart based on work of John R. Christy, University of Alabama, Huntsville
Model output: KNMI Climate Explorer

Source: https://judithcurry.com/2015/12/17/climate-models-versus-climate-reality/

Figure 5-7 shows observations of recent atmospheric temperatures made by multiple sources, including the average of three satellite record sets and the average of four radiosonde record sets. These observations yield dramatically cooler values than those projected (red curve) by the average of the prosecution's latest generation of theory-based greenhouse gas climate change simulation models (CMIP5).[85]

When rigorously developed models go wrong, the underlying theory upon which those models are developed must be flawed (see "When Computer Models Get It Wrong, Question Their Basis"). Except, apparently, when those models are developed in support of the prosecution's theory of a crime.

Rather than admit its theory is flawed, the prosecution prefers to take the flawed theory's model projections, wave them hysterically at the jury, and proclaim a climate change Armageddon is just around the corner!

From: *Climate Models versus Climate Reality*[85]:

> Rain and snow are largely dependent upon the temperature difference between the surface and the mid-troposphere. When there's little difference, air in the lower atmosphere does not rise, meaning that the vertical motion required to form a cloud is absent. When the difference is large, moisture-laden surface air is very buoyant and can result in intense rain events.

> Getting the difference systematically wrong in a climate model means getting the rainfall wrong, which pretty much invalidates regional temperature forecasts…

> Indeed, the models have this temperature differential dead wrong. Over the period of study, they say it should be remaining the same. But, in fact, it is growing in the real world, at a rate nine times what is predicted by the models over this study period.

Jurors should contemplate the prosecution's bizarre situation.

It is in *complete denial* of the mountain of evidence contradicting its climate change theory of a crime while at the same time the only evidence supporting its dogma is the product of computer simulation models based on the same unvalidated theory contradicted by real-world evidence!

Given a choice between solid real-world evidence and dubious projections of theory-based models whose underlying theory is unvalidated, the prosecution doubles down on its dubious projections and ignores the evidence! That's not science, it's dogmatic faith in contradicted theory.

More Missing Tropical Tropospheric Greenhouse Gas Warming

Figure 5-8 offers jurors additional compelling evidence contradicting the climate change theory on the basis of the missing greenhouse gas tropical tropospheric warming. The original chart is from *An Overview to Get Things into Perspective*[68] at Ole Humlum's excellent website Climate4You. com.[86]

The blue and red charts compare histories of lower tropospheric temperature anomaly with the average outgoing longwave radiation (IR, Earth shedding heat) for each year (January 1979 through January 2001) over a tropical region (20°S to 20°N latitude). Recall that outgoing IR is the claimed driver of the greenhouse gas theory's climate warming process and that the prosecution's greenhouse gas climate change theory requires a distinctive and strong greenhouse gas warming hot spot to be present in the tropical troposphere.

If the prosecution's climate change theory has any merit, these charts should reveal a clear and obvious relationship between outgoing IR (in red, the claimed driver of greenhouse gas warming) and temperature anomalies (in blue).

According to greenhouse gas theory, the red curve (outgoing IR) is driving changes in the blue curve (tropical lower tropospheric temperature anomaly). But clearly, the relationship between the red and blue curves is tenuous at best and certainly *does not show* the red *driving* the blue!

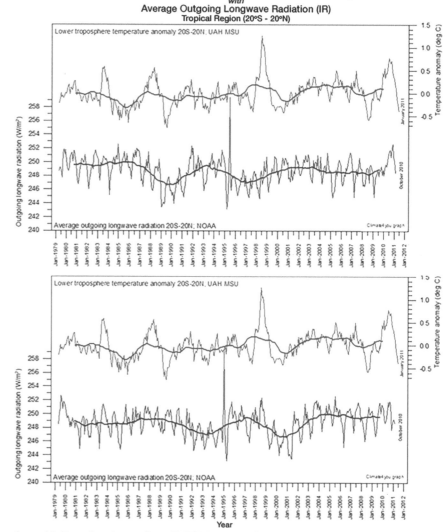

Figure 5-8
Which Chart is Correct - Top or Bottom?
Comparing Lower Troposphere Temperature Anomaly
with
Average Outgoing Longwave Radiation (IR)
Tropical Region (20°S - 20°N)

Source: http://www.climate4you.com/images/OLR%20Equator%20NOAA%20and%20UAH%20MSU%20since%201979.gif

One of the two red graphs is correct while the other red graph is flipped left to right. Can jurors tell with absolute certainty which chart is correctly oriented?

Is that missing greenhouse gas hot spot evident?

There is no consistent relationship between peaks and valleys of either set of curves. If there were a strong relationship that clearly showed annual changes in outgoing IR produce corresponding annual changes in temperature anomalies (in accordance with the prosecution's climate change theory), then the jury could quickly recognize which chart is the actual chart and which chart has its red curve flipped horizontally (right and left sides switched).

Because greenhouse gas theory claims that virtually instantaneous radiant heat transfer from Earth's surface to the atmosphere and then back to Earth's surface is driving climate warming (the so-called "trapped heat"), the theorized heat transfer must be virtually instantaneous (because all radiation travels at the speed of light) and radiant heat transfer between Earth's surface and its atmosphere should occur virtually without delay.

If the prosecution's climate change theory were valid, then over the course of a year the atmospheric response to changes in outgoing IR would be clear and evident. Yet neither curve shows the annual temperature anomaly change (blue) consistently reflecting annual outgoing IR change (red). Clearly, there is no consistent relationship, yet one is *claimed to be the driver of the other!*

As Figure 5-8 shows beyond any reasonable doubt, between 1979 and 2002 the persistent growth of atmospheric CO_2 produced no observed consistent reaction to outgoing IR that justifies any claim atmospheric CO_2 growth is a detectable climate change force.

Polar Amplification, Another "Must Be" of Climate Change Theory That Isn't!

Another of the mandatory consequences of the prosecution's climate change theory is polar amplification of CO_2-driven warming, a theory-based requirement that polar regions warm first and most dramatically.

From one of the IPCC's working group reports for its Fifth Assessment Report (AR5):[87]

> Polar Amplification – zonal mean surface temperature warming at high latitudes exceeds global average temperature change.
>
> New temperature reconstructions and simulations of past climates show with high confidence polar amplification in response to changes in atmospheric CO_2 concentration. In the absence of a strong reduction in the Atlantic Meridional Overturning, the Arctic region is projected to warm most (very high confidence).

In short, if the prosecution's climate change theory is valid, then polar regions will show evidence of the theorized warming before it is observed in nonpolar regions.

But what do recent studies of this effect reveal?

In an April 2018 article titled "In 2015, Climate Scientists Wrecked Their Own CO_2-Forced 'Polar Amplification' Narrative,"[88] compelling testimony is given by several papers that directly contradict the prosecution's polar amplification effect.

One of those papers, "How Increasing CO_2 Leads to an Increased Negative Greenhouse Effect in Antarctica,"[89] published in *Geophysical Research Letters*, concluded:

> We can conclude that the role of CO_2 in the Antarctic climate is somewhat different to the rest of the planet: Increasing CO_2 has a rather small direct effect on the Antarctic climate; it even tends to cool the Earth-atmosphere system of the Antarctic Plateau. The analysis…did not result in any statistically significant surface temperature trend on the East Antarctic Plateau during the last decades. They even found a slight (but statistically not significant) cooling trend for the centre of Antarctica.
>
> We have compared…surface measurements of broadband LW [longwave, IR] upward fluxes from South Pole with model estimates of this quantity compiled for the fifth IPCC assessment report (fifth phase of the Coupled Model Intercomparison Project), analogue to the comparisons reported by Wild et al. [2012]. This comparison shows that GCMs [General Circulation Models] tend to overestimate the surface temperature.

This research confirms that the prosecution's theory-based global climate change simulation models consistently overestimate temperature projections. In other words, the models *consistently* run hot, indicating either (1) the theory is flawed or (2) every one of the models based on that theory has a significant error that causes its projections to be too "hot" (projected temperatures are too high), a highly unlikely alternative. For more about hot-running models, later in this chapter read the selection titled "When Computer Models Get It Wrong, Question the Theory Modeled."

In April 2018, Australia's Joanne Nova posted an article, "Climate Change Means Greenland Is the Same Temperature Now as 1880,"[90] in which she noted:

> Greenland hasn't been showing signs of warming since man made CO_2 [emissions] started rapidly rising after World War II. Indeed Greenland has [not been] responding to CO_2 for 140 years or maybe a million [years].
>
> Serious researchers have known this for years. It's not like a flat trend suddenly popped up to surprise us.[90]

According to the prosecution's climate change theory (as restated in the IPCC AR5), one of the key consequences of polar amplification is that:

> …the Arctic region is projected to warm most (very high confidence).

So the prosecution's *theory* requires stronger polar (Arctic and Antarctic) warming than what is observed in any other region (temperate, subtropical, or tropical). Is there any evidence of the prosecution's theory-required pronounced polar warming?

Figure 5-9 shows 160 years of temperature anomalies in Greenland. Shaded areas between the anomalies and the zero of the temperature scale indicate cooling or warming.

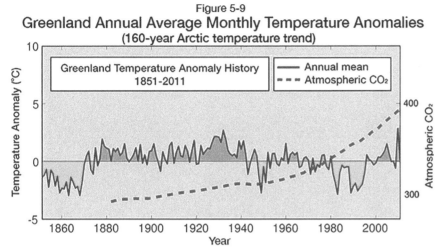

Figure 5-9
Greenland Annual Average Monthly Temperature Anomalies
(160-year Arctic temperature trend)

Based on: Figure 2 from Supplement to "Influence of temperature fluctuations on equilibrium ice sheet volume" by Troels Bogeholm Mikkelsen, et al.

The zero of the temperature scale separates warming anomalies from cooling anomalies (greater or lesser temperatures than the long-term average). Warming above the long-term average occurs whenever the curve is positive (above the zero line, *shaded red*) and cooling below the long-term average occurs whenever the curve is negative (below the zero line, *shaded cyan*).

Atmospheric CO_2 (dashed curve) is overlaid for dates beginning with 1880. Note the lack of *any* theory-consistent relationship between growing atmospheric CO_2 and Greenland's climate (temperature) change.

During the most recent 31 years (from 1980 to 2011), there is greater evidence of cooling than warming. While there was a spike in 2010, the anomaly for 2011 was virtually zero (at the extreme right and difficult to see on this chart).

Notice between 1980 and 2000 the theory-inconsistent behavior of atmospheric CO_2 and temperature anomaly. No consistent relationship between atmospheric CO_2 and temperature anomaly is evident, indicating that, contrary to the prosecution's theory, there is no meaningful relationship between the two.

These real-world observations clearly show another bedrock requirement of the prosecution's theory, polar amplification, simply isn't happening. This conclusion is consistent with Figure 4-23 (page 89) showing Greenland ice core records for the Holocene interglacial that similarly defy the prosecution's climate change theory.

Summary: When Inconvenient Records Contradict Greenhouse Gas Theory

1. Both satellite and radiosonde (weather balloon) observations show no unusual climate warming that can be linked to atmospheric CO_2 during the 40 years from 1979 to 2018 (inclusive).
2. A small upward shift in temperature, possibly related to the strong 1998 El Niño, is evident in both the satellite and radiosonde record. This baseline temperature shift is not a consistent change relating to atmospheric CO_2 change.
3. The prosecution's climate change theory-based models project a *theory-required* strong warming "hot spot" in the tropical (+/- 30° latitude) mid-troposphere (8 km to 12 km altitude). Yet no evidence of this theory-required hot spot exists. In fact, *slight cooling is observed* in that region where greenhouse gas theory demands evidence of strong warming! When theory disagrees with observation, the theory is invalidated and *must be* rejected.
4. Prosecution theory requires (and theory-based models project) warming in polar regions will be more pronounced than in lower latitudes. This polar amplification of greenhouse

warming is a robust requirement rooted in the prosecution's greenhouse gas climate change theory. Yet once again, measurements show a *failure of the required polar amplification* to materialize. Some evidence indicates Antarctic polar regions are actually cooling. Yet another example of real-world observations that contradict the prosecution's greenhouse gas climate change theory!

5. In accordance with *the scientific method*, the prosecution's theory and theory-based models are invalidated by their failure to agree with observations.

Is Evidence of Man-Made Climate Change Really Man-Made Evidence of Climate Change?

The prosecution's case against fossil fuel is rooted in its claim that fossil fuel combustion produces CO_2 emissions that are the source of atmospheric CO_2 growth that is said to be responsible for global climate warming.

If so inclined, jurors can explore challenges to the prosecution's theory by reading "SURFRAD Data Falsifies the 'Greenhouse Effect' Hypothesis" by Carl Brehmer and other papers coming to similar conclusions.[55–62]

A record of steadily growing atmospheric CO_2 is well established, increasing by 98 ppm during the 100 years ending in 2014. Absent a mitigating superior climate change force, if the prosecution's theory were valid, a consequent global climate warming should be observed in accordance with the theorized response to that century-long atmospheric CO_2 growth (most of which occurred in the last 71 of those years).

As if to punctuate the problems with the prosecution's theory, in the first decade of the new century, global warming took a vacation when the great "pause" began. Based on the newly-emerging USCRN ground station records, it appears that the pause has continued for two decades through 2019. The extended pause provides ample evidence of the fallacy of the prosecution's climate change theory, stripping any legitimacy from the prosecution's claims while revealing global warming's emperor (climate change theory) to be stark naked!

As this inconvenient warming pause marched on toward its first decade, the term *global warming* had to quietly be relabeled *climate change* to avoid the awkward embarrassment of demanding action to fix something that wasn't happening! Among the prosecution's top priorities was the need to quickly hide the exculpatory evidence (the pause) and then claim it never existed!

Homogenization

A number of well-known problems with legacy temperature recording stations[78] compromise the quality of a significant number of land-based temperature records, particularly those in urban areas. Perhaps the most widespread problem is the one known as the "urban heat island" effect. Temperature recording devices that had been sited in or near rural towns during the 19th and early 20th centuries have had surrounding areas significantly transformed by population growth and expansion of built-up areas that creates artificial heating sources and sinks that compromise the quality of records as the city or town expands around these temperature recording devices.

To address this issue and others, NOAA took two actions. They created the United States Climate Reference Network (USCRN)[175] of 114 advanced, well-sited climate monitoring stations across the contiguous 48 states. In January 2005 the USCRN began recording quality temperature data.

NOAA also developed a process known as *homogenization*, ostensibly to correct suspect temperature records (e.g., the urban heat island effect) by attempting to alter them to what they would be if not for the changing environment that has biased temperature readings over portions of the past 150 years. But that isn't what happened.

A few observations about the wisdom of correcting records:

- Given the large number of global surface stations recording temperature, homogenization is a monumental task. A simple methodology error could throw off the entire record.
- Each recording station is unique; standardized "one size fits all" corrections are often inappropriate.
- To engage in temperature homogenization of raw temperature records at a time when those records might be in great demand by researchers is a risky proposition. It would be just too easy to tweak the process to create the appearance of support for climate change theory where none exists, particularly when the effort is undertaken by those who support climate change theory and feel compelled to defend it.
- The record as recorded is better than manipulated records, no matter how honorably the manipulation might be motivated or how carefully performed. An appropriate history associated with each station would identify known defects in the original temperature records. Armed with this information, researchers could use the information accordingly, making their own adjustments as required.
- Each reporting station has a history of site changes and known defects. There is no real need for the original records to be adjusted.

Nevertheless, throwing caution to the wind, in the latter part of the first decade of the 21st century, temperature record homogenization was initiated, and it continues today to some extent both in the USA and internationally.

Abundant evidence exists that after 2008, when the hiatus (or pause) in global temperature warming neared a full decade, a number of long-established international temperature records were quietly altered by this process known as "homogenization".

The resulting changes appear to justify the worst concern identified in the third bullet above. There is compelling evidence homogenization is the source of considerable man-made climate change because both the untampered evidence and the new USCRN temperature record from 2005 through 2019 unequivocally reject the prosecution's climate change theory whereas some homogenized records conveniently appear to support it.[91–94]

At best, legacy temperature records are best estimates with ground station sites exhibiting varying degrees of accuracy; at worst, the legacy temperature record is corrupted by efforts to make it consistent with flawed climate change theory as represented by model projections.

Despite these shortcomings, the legacy surface temperature records are the best available long-term observational evidence particularly when they haven't been the object of man-made climate change in the form of homogenization.

Given the evidence that homogenization was crafted to either create the illusion of recent climate warming where none had existed or create the illusion that well-recorded warming in the past *hadn't* occurred (cooling the past),[95] it is not surprising to discover that, subsequent to homogenizing, the warming pause in some ground station records appears to have disappeared—homogenized out of existence!

Even more creative, using their magic wand of homogenization, the prosecution's agents "cooled" the 1930s, formerly the hottest decade on record, and "warmed" the 33-year cooling trend (1941–1974) that had raised concerns of a new ice age:

> Meanwhile, the National Oceanic and Atmospheric Administration has been conducting highly suspicious temperature data manipulation. The changes in the temperature data consistently make the past seem cooler, which in turn makes the present seem warmer.[96]

> NASA and NOAA dramatically altered US climate history, making the past much colder and the present much warmer... NASA cooled 1934 and warmed 1998, to make 1998 the hottest year in US history instead of 1934. This alteration turned a long term cooling trend since 1930 into a warming trend.[97] [from 1930 to 1998]

Deserving of the *Winston Smith*[98] *Award* for altering reality in an attempt to create *the illusion* that temperature records are in compliance with the prosecution's invalid climate change theory, these actions have put a real twist to *the scientific method.*

When the observed temperature record contradicts theory, well then, it isn't *the theory* that must be altered...no, no, no...why, it's those pesky *original temperature records* that must be altered!

Members of the jury are told they mustn't believe their own senses or satellite readings or radiosonde measurements or historic unaltered temperature records. Not because they are flawed, but because they do not conform with the prosecution's theory.

When climate change theory becomes climate change dogma, it loses any connection with science.

On the basis of homogenized temperature records, the prosecution routinely proclaims "the hottest year on record," often citing the altered records as the basis for its proof.

A few examples follow.

Those living in the eastern US might recall the bitter cold period from late December 2017 into mid-January 2018. Anyone who lives in the northeast US will need to keep a sharp memory because NOAA's homogenization artists have been diligently rewriting weather history.

James Delingpole (*NOAA Caught Adjusting Big Freeze out of Existence*[99]) discovered that the inconvenient bitter cold winter of 2017–2018 in the northeastern US was being systemically erased by NOAA's homogenization of the temperature record.

Delingpole said, "the recent record-breaking cold across the northeastern US...caused sharks to freeze in the ocean and iguanas to drop out of trees." Yet, according to NOAA's *revised temperatures*, January 2018 temperatures in the northeastern US were virtually indistinguishable from the 20th-century mean, and much colder January temperatures were routinely experienced over the past 100 years!

In another investigation of NOAA's peculiar activity, Paul Homewood recently discovered NOAA "cooking the books" to erase the bitter cold winter of 2013–2014 in New York State's Central Lakes region.

Here are some quotes from NOAA's own summary of the November 2013 to March 2014 five months of winter:

> Buffalo…finished with an average temperature of 26.4 degrees Fahrenheit. This is the 8th coldest alltime [sic] and coldest since records were moved to the airport in the summer of 1943.

> Rochester…tied for [its] 13th coldest alltime [sic], and coldest since the winter of 1993–94…

> Watertown…was tied with 2003–04 for 4th coldest. The average temperature of 23.1 degrees Fahrenheit …was only 1.1 degrees warmer than the all-time coldest 5-month stretch (1977–78 at 22.0F)."

Comparing archived January records for 1943 with 2014, Homewood discovered that temperatures (of the stations that existed in both years) were actually 2.7°F colder in 2014 than they were in 1943.

Yet NOAA's "adjusted" (homogenized) temperatures substantially *chilled* the 1943 figures to reduce the differences, making 2014's temperatures appear to be just 0.9°F colder than they were in 1943.[100]

That's a whopping 1.8°F of *man-made climate change*, taxpayer-funded fraud designed to deceive the very taxpayers who paid for it!

Furthermore, because of the change of use at Syracuse airport (which today is far busier), that station *should not be used* for climate purposes (as it is subject to a strong urban heat island effect—blacktop runways, jet exhaust, etc.).

Dropping out the challenged Syracuse airport station, the actual average temperature difference in the Central Lakes region of New York State *should show that region's 2014 temperatures to be 3.3°F colder than they were in 1943!*

Instead, NOAA reports the difference as just 0.9°F colder. They managed this by cooling the 1943 data. That's not *science*, NOAA, it's sloppy work at best, fraud at worst.

In August 2017, Tony Heller published a comprehensive exposé[101] concerning NOAA's climate change activities doctoring the US temperature records that clearly show the extensive systematic activities engaged in by NOAA to manipulate the historic US temperature records. All this to hide the inconvenient truth that the historic legacy temperature record doesn't support a claim of dangerous or unprecedented climate warming.

From Heller's study, "In 1989, NOAA Reported No US Warming…":

> Washington, Jan. 25—After examining climate data extending back nearly 100 years, a team of Government scientists has concluded that there has been no significant change in average temperatures or rainfall in the United States over that entire period.

> The study, made by scientists for the National Oceanic and Atmospheric Administration… is based on temperature and precipitation readings… from 1895 to 1987. (*The New York Times*, January 26, 1989)

Using US Historical Climatology Network (USHCN) data (Figure 5-10), Heller shows the contrast between original temperature readings (blue, empirical data) as reported by weather stations and what is now reported (red) as the temperature record over that same time. Clearly, there is evidence of man-made climate change, but not the prosecution's man-made climate change claimed to be caused by atmospheric CO_2. This man-made climate change is created by the necessity to manufacture the appearance of support for an invalid theory.

Several generations of students have been taught this invalid climate change theory as if it were gospel and the prosecution must tap that reservoir of true believers while it still believes.

Some might claim that temperature measurements were taken under different conditions and at different times of day, but that is an excuse for homogenization, not a legitimate basis for actually altering raw temperature readings of selected data sets by estimating what one thinks they would have been if measured under different conditions! The data and conditions under which readings were obtained are part of the full record. There is no legitimate purpose served by tampering with that evidence.

But that is exactly what has been done and continues to be done. Figure 5-10 illustrates the degree to which warm temperatures in the past (blue curve) have systematically been reduced (red curve), while more recent temperatures are raised to *manufacture an illusion of dramatic warming* over the past 100 years. All for the sole purpose of confusing the jury with tainted evidence!

These man-made climate alterations amount to prosecutorial tampering with the evidence in the name of data homogenization!

Viewing Figure 5-10, the jury should ponder this question:

> Would jurors believe 20th-century temperature warming is a significant problem if the untampered measurements (in blue) were the basis for their determination?

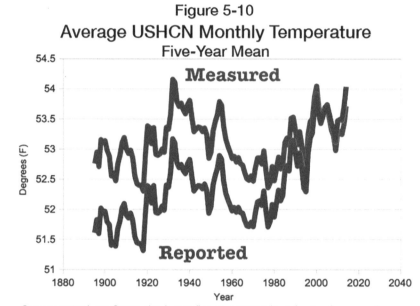

Figure 5-10
Average USHCN Monthly Temperature
Five-Year Mean

Source: "100% of US Warming is Fake" by Tony Heller (realclimatescience.com)

Such fraud coming from historically ethical agencies (NOAA, USHCN, NASA GISS) would have been unheard of prior to the political pressure to manufacture this deceit. After 2008, these science agencies were politically encouraged to engage in a radical increase in fakery designed to support the prosecution's climate change theory. All, of course, in the interests of the public welfare.

Noble motivation + ignoble methods = fraud.

Right out of Orwell's *1984*, a handful of *Winston Smiths*[98] began systematically altering the measured records to support the political objectives of politicians with an agenda. A better example of prosecutorial evidence-tampering may not exist.

As Heller's study concludes, "This is the biggest and most cynical scam in science history."

While this is ongoing, what does the jury hear from the prosecution's parrots in "big media"? An exposé? Outrage? Nope, just "crickets" as the news media issues a collective yawn to fill the gaps between scary narratives claiming yet another climate disaster caused by—you guessed it—fossil fuels. Scary fake news stories sell, selling yields greater profits, and truth suffers at the altar of political expediency.

As Figure 5-11 illustrates, the US isn't the only government whose agencies are involved with systematic alteration of established temperature records. The Australian Bureau of Meteorology is performing the same man-made climate change magic in Australia.

Like their counterparts in the US, these adjustments (appropriately termed "cooking the books" by some), have been suspiciously consistent at *cooling past* temperatures while *warming more recent* temperatures.[102] The probability of honest adjustments persistently behaving that way is virtually nil since every temperature measurement station is unique.

NOAA's alterations have also expediently served to remove the US's historic "dust bowl" heat waves of the 1930s, which decade (ironically) still holds the record (by far) for most US state all-time high temperature records set in any decade (see more below in "Do US Climate Records Support a Claim of Significant Recent Climate Warming?").

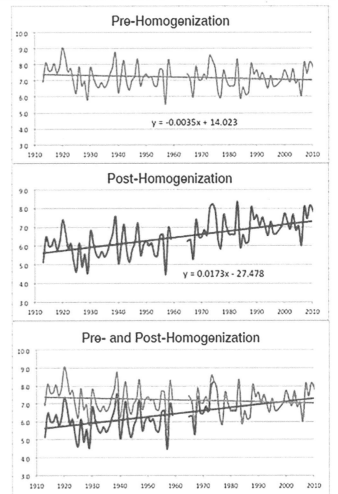

Figure 5-11

Rutherglen, Victoria, Annual Average Minimum Temperatures
Blue is pre-homogenization showing cooling of 0.35° Celsius per century.
Red is post-homogenization showing warming of 1.73° Celsius per century.
Clearly, homogenization changes the temperature trend very dramatically.

Source: "Homogenization of Temperature Data By the Bureau of Meteorology" (Australia), Reference 7, Brendan Godwin, December 2016 (https://wattsupwiththat.com/2016/12/21/homogenization-of-temperature-data-by-the-bureau-of-meteorology/)

Other alterations removed the dramatic cooling of the mid to late 20th century that had brought fears of a new ice age. Still other alterations raised more recent temperatures to heighten the warming illusion. Fortunately, the records[7] used as the basis for Chapter 4's global average temperatures came with no warning that they had been homogenized and, in any event, the story they tell in Chapter 4 strongly contradicts the prosecution's climate change theory.

Were all these data manipulation shenanigans begun in order to claim the hiatus (or "pause") in global warming over the past two decades never existed? Much to the prosecution's chagrin, the hiatus is, to this day, clearly visible in nearly 20 years of radiosonde (weather balloon), satellite measurements, and the new USCRN records.

These after-the-fact climate change manipulations would have made George Orwell's *Winston Smith* (*1984*)[103] proud.

The prosecution's calculated alteration of historic climate records bear a striking resemblance to the infamous "Hockey Stick" deception, referred to by some as "Mann-made" climate change.[104]

That scandalous effort attempted to eradicate the most well-established significant global climate variations of the past 2,000 years (the Roman and Medieval Warm Periods and the Little Ice Age).

Evidently, because those well-established global climate variations just didn't fit the prosecution's climate change narrative, they simply had to be eliminated by any means, legitimate or otherwise.

> A major person working in the area of climate change and global warming sent me an astonishing email that said, "*We have to get rid of the Medieval Warm Period.*" (David Deming, 2005)

Think about that statement. It exposes the fact that the prosecution *knew* that the evidence contained in well-established climate records could testify to the frivolity of the prosecution's climate change theory and exonerate atmospheric CO_2.

The only "solution" deemed acceptable was to *create an illusion* that the well-established Medieval Warm Period and Little Ice Age had never occurred!

Apparently, those engaged in temperature data homogenization believe if either historic or recently observed temperatures do not support climate change theory, then it isn't the theory that must change, it's the record that must change!

Gone is the record heat and drought of the 1930s[105]…poof, homogenized right out of existence! The 1930s dust bowl? What 1930s dust bowl? Mid-20th century cooling? Gone. Just as in Orwell's classic *1984*, historic records have simply been changed so *they appear to match* the prosecution's climate change *theory*.

From one of the investigations[94] exposing these temperature homogenizations:

> In…large cities where there are tall buildings, it has been well known…that these temperatures would be half to one-and-a-half degrees C cooler if the tall building and the city wasn't there. It is a phenomenon known as "the [urban heat] island effect"…

> But the [problem] is that [those making the adjustments are] adjusting [surrounding] country temperatures upwards by half to one-and-a-half degrees C so that they match the city tem-

peratures[!!]. That's creating about a degree C of warming when if they had adjusted the city temperatures down half to one and a half degrees C, they'd be creating approximately a half a degree of cooling.

Got that? Scientific fraud being perpetrated on a grand scale by homogenizers of the data. In order to compensate for the impact of the urban heat island effect on the temperature record, rather than properly compensate by adjusting artificially high urban temperatures *downward*, the more accurate rural temperature records in outlying areas are being adjusted *upward*. Such adjustments make temperatures inaccurate while expediently *expanding* the urban heat island effect rather than *eliminating* it. This scheme conveniently creates evidence of "global warming" out of cooler air! If the records don't match theory, no problem. Just alter the records so they do!

Such data tampering is both inexcusable and fraudulent. It serves only to increase *the illusion* of global warming.

If the jury were to look for a single word to characterize homogenization, look up *Orwellian* and *fraud* and make the difficult choice between the two. Better yet, combine them into the perfect phrase: "Orwellian fraud." A better example of prosecutorial misconduct would be difficult to find.

What may have been begun as an attempt to compensate some temperature records for measurement siting issues clearly became a shameless perversion of *the scientific method* for political purposes. At the root of all this deception is a concerted effort to dupe the jury.

Homogenization and its widespread costly impact (e.g., *Paris Agreement*) and the attendant corruption of science in education have become symbols of the greatest scientific fraud in the history of sci-

ence. The jury should consider the extent of this fraud and the devastating impact it is having on science, honesty, integrity, and public faith in government institutions. Truth must surmount politics and put an end to this fraud.

Summary: Is *Evidence of Man-Made Climate Change* Really *Man-Made Evidence of Climate Change?*

1. Is evidence of man-made climate change really man-made evidence of climate change? Evidently. When historic temperature records contradict the prosecution's climate change theory, the records have been subject to questionable record-tampering known as *homogenization* that generally creates the illusion that homogenized records appear to support the climate change theory! Such manipulation of raw empirical evidence is particularly inappropriate should the intent have been to deceive the jury with tainted evidence. There is no useful scientific purpose served by ongoing data manipulation that conveniently adjusts recorded temperatures by chilling past records while warming recent records to create the illusion that the prosecution's theory of a crime might be valid. It is highly unlikely such a coincidence of fortuitous adjustments occur by chance.

2. In response to the nearly two-decade warming hiatus that began around the turn of the century, the phrase *global warming* was replaced with *climate change*, meaning, of course, *human-caused climate change*. The prosecution found it just too embarrassing to prosecute a trial against carbon dioxide and fossil fuels for the crime of creating global warming when little or no significant warming was evident. Much easier to attack a more nebulous "climate change" that requires at least thirty years to establish and can mean almost anything from dramatic cooling to dramatic warming.

3. Faced with the evidence of temperature data that contradict its theory of a crime, rather than reconsidering its flawed theory, the prosecution's agents have tampered with the evidence in an apparent attempt to create the illusion that the records support the prosecution's indictment of CO_2! If the prosecution's theory didn't match the evidence, well then, the evidence simply had to be altered to create *the illusion it did match* theory. Past warm temperatures are cooled while more recent cooler temperatures are warmed. Presto, man-made evidence of climate warming!

4. By homogenizing the evidence and manufacturing novel new claims about characteristics of CO_2 emissions, the prosecution's advocates turn *the scientific method* on its head.

Do Historic US Climate Records Support a Claim of Significant Recent Climate Warming?

The 15-Year USCRN Record (2005-2019)

Records for surface station temperature measurements dating to the 19th century are known to have significant siting issues[78] (urban heat island effect, proximity to hot exhaust fans and blacktop surfaces, etc.). During the latter part of the 20th century these issues skewed the temperature measurements for growing cities, frequently recording warmer measurements than would have been recorded in the absence of development.

NOAA addressed this human-caused problem with its temperature records by creating the US Climate Reference Network (USCRN) that uniformly distributed 114 new carefully-sited and well-maintained temperature-reporting stations across the contiguous 48 US states. These new stations began monitoring in 2005 and have amassed 15 years of uncorrupted records (14 years of temperature change).

Figure 5-12 plots the average of the 114 station monthly temperature anomalies for each month from January 2005 through 2019. This graph plots anomalies, which are the difference between the actual measured temperatures and a particular average temperature (the exaggerated "tips" of a temperature graph).

Temperature anomalies clearly show the developing warmth from the 2015-2016 strong El Niño event as well as the subsequent cooling trend. Because anomalies represent the tips of the actual temperature graph, the temperature change between any two dates is simply the difference between the anomalies for those two dates. For example, subtracting the anomaly for January 2005 from January 2010's is the same as subtracting the January 2005 temperature from January 2010's temperature. Figure 5-12 shows *no significant warming trend* and *slight cooling* during the first fifteen years of monitoring these new USCRN climate stations (January 2005 through December 2019).

Monthly temperatures for the 114 USCRN sites are averaged to create a monthly temperature profile for each year. The net monthly temperature change from 2005 is averaged over all twelve months to produce a net temperature change since 2005. Based on the most current record through 2019, the 14 year aggregate weighted monthly change reveals temperatures *cooled* by -0.41°C (-0.74°F) since 2005 while atmospheric CO_2 *grew* by 30 ppm, again, *contradicting* the demands of greenhouse gas climate change theory.

In Defense Exhibit B, jurors will find a section "More About USCRN Records" that contains a more in-depth look at the evidence collected by these 114 new climate monitoring stations from January 2005 through 2019. The jury will view compelling evidence that since these significantly improved monitoring stations have come online a *net cooling* of the 48 contiguous US states has been recorded.

Figure 5-12
NOAA/USCRN Contiguous U.S. Monthly Average Temperature Anomaly
(Fifteen years, 2005 through 2019)

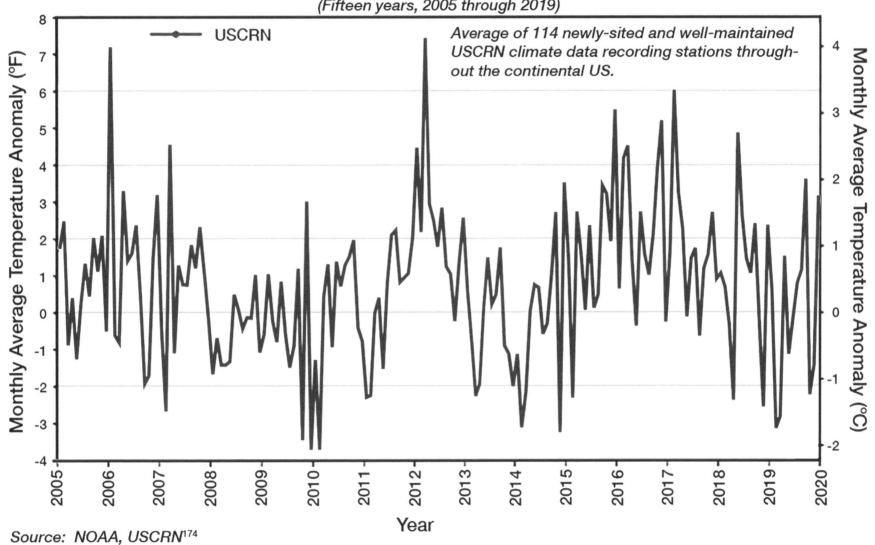

Average of 114 newly-sited and well-maintained USCRN climate data recording stations throughout the continental US.

Source: NOAA, USCRN[174]

The 2005 through 2019 USCRN records strongly support the lack of any observed non-trivial warming trend across the continental USA during the first two decades of the 21st century.

Yet the prosecution claims recent years are among the hottest on record!

So who are the *real* climate deniers?

The prosecution, who clings to its climate change theory while consistently denying real-world evidence?

Or the defense, who offers the jury compelling real-world evidence that consistently contradicts the prosecution's theory?

Hottest Year on Record—What Does That Really Mean?

It is fairly routine to hear claims that recent years are among the hottest years on record.

Yet consider the current climate scenario. Naturally warming climate since the end of the Little Ice Age features alternating natural multi-decade periods of rising and falling temperatures. Because natural climate change tends to raise temperatures slowly as climate changes from cool period to warm period, typically a considerable time is required for this slow change to be completed.

Superimposed over shorter decadal warm-cool cycles, shouldn't a long-term warming trend *be expected* as Earth's climate emerges from the LIA? Why should it be the least bit remarkable to experience periods when temperatures are the highest on record when the earliest records only began in the latter decades of the recent very cold Little Ice Age?

Look at Figure 3-4 (page 43) showing solar activity trends during the Medieval Warm Period, the Little Ice Age, and post-LIA warming. During the years 1050 to 1220 of the MWP, the years 1660 to 1780 of the LIA, and the post LIA years 1880 to 2000, it would be highly likely that climate warming would be fairly consistent, save for a few perturbations around the long-term climate trends. Had records been maintained during those periods the jury should expect there to have been many times that qualified as the hottest year on record.

The point is, climate change is nothing new. It is in response to a variety of factors, and there is no evidence to definitively link any climate change to fossil fuel use or any other possible cause of atmospheric CO_2 growth.

Remember, just because temperatures might be rising doesn't mean they are rising for the reasons claimed by the prosecution.

In Chapter 4, the jury examined the validity of the prosecution's key allegation relating climate change to changing atmospheric CO_2. The jury discovered the testimony of real-world records unequivocally shows the prosecution's theory is invalid nonsense.

It is very clear that no real-world evidence exists to support the prosecution's allegations against fossil fuel use, the claimed source for the 20th-century growth of atmospheric CO_2 that, in turn, is blamed for observed climate change.

Naturally, during a rebounding climate from a cold period (the Little Ice Age), *there will always be those years* that can safely be proclaimed to be the warmest on record—until, of course, the next warmest on record as temperatures continue to creep upwards naturally! This is demonstrably true when the record goes back no further than the century during which the LIA is said to have ended!

What is odd or peculiar about such natural warming? Nothing. Are those who find it unusual denying natural climate change? Are they the real deniers? Just how long have temperature records been kept? In terms of Earth's long climate history, in the blink of an eye. Not long enough to know much about what should be expected of climate change or when to expect it.

Note the red block on the time scale of Figure 4-23 (page 89). Is the recent climate typical of Holocene interglacial climate? The current climate is clearly consistent with Holocene climate and not at all unusual.

Looking at Figure 4-23 again, is the current climate warmer or colder than most Holocene interglacial climate? Based on Greenland's ice core temperature history, current climate is *much cooler* than most climate during the Holocene interglacial.

Nevertheless, frenzied proclamations of "record heat" are delivered with the implicit suggestion such warming *must be* unnatural and *must be* caused by CO_2 growth that *must be* the result of humans using fossil fuel for energy!

As the evidence in Chapter 4 unequivocally demonstrates, not only are such claims unwarranted, they are simply another attempt by the prosecution to mislead the jury to try to justify its flawed theory!

Since the past century of climate change is perfectly normal for *natural* Holocene climate, jurors might reasonably question the prosecution's motive for its evidence-contradicting theory-based prosecution of atmospheric CO_2 and fossil fuels.

Until scientists fully understand the forces that lead to natural climate change, it will be impossible to accurately attribute any portion of recent observed change to any unnatural force (e.g., fossil fuel use).

Natural climate change science remains poorly understood and records certify there is simply no evidence-based relationship between climate change and changing atmospheric CO_2. Just because the past 130 years saw both atmospheric CO_2 and global temperature increase does not constitute sufficient evidence the two have a causal relationship. Those who suggest otherwise are trying to deceive the jury. (See correlation coefficient in the glossary).

There are a few things we know for certain. One of them is that climate changes. Naturally. All the time.

Considering the relatively brief human life span, our experience with climate change cannot inform us of the extent and relative frequency of routine natural climate change.

Fortunately, records can be examined to see if they support the prosecution's claims of recent dramatic global climate warming. One such indicator is the decade during which all-time temperature extremes (highest and lowest temperatures) were recorded for each state in the USA.

Questions to be examined with these records include:

- Do extreme temperature records for US states dating back to the late 19th century provide any evidence Earth is experiencing unusual climate warming at this time?
- Is there any historic evidence in the period since 1950, often cited by the prosecution as providing support for its theory?
- What about evidence since the beginning of the 21st century?
- If climate has experienced "unprecedented" warming as alleged, is it reasonable to expect that new all-time state high temperature records should be being set in more recent years?

- Do US state all-time high temperature records support claims that recent years are among the hottest on record?

What are jurors to think about these virtually ignored reports of cooling in 2018 and 2019?

- Plummeting summer temperatures in the US,[106]
- US had its coldest April in more than twenty years,[107]
- Climate change means Greenland is the same temperature now as 1880,[69]
- Arctic sea ice soars, polar bears start hunt early—second year in a row.[108]
- On January 31, 2019, the State of Illinois recorded its all-time record low temperature of -39°C (-38°F).[109]

Did jurors see or hear about any of those reports in the news? If not, could it be that all these reports run contrary to the climate change narrative routinely inserted into news stories? Should inconvenient records be ignored as politically incorrect or just *homogenized* away?

Again, who are the real climate deniers?

Prior to homogenization of temperature records, the 1930s had been known as the hottest recorded decade in US history. By tampering with temperature records, the prosecution is trying to fool the jury by forcing the altered record to match its theory and appear to justify its claim the second decade of the 21st century is the hottest on record. If the prosecution were so certain of its theory's validity, it wouldn't need to alter the temperature records that contradict its theory.

Table 5-1 summarizes the all-time high and low temperature records[109] for each of the fifty US states. Study this table for a moment. Note the inconsistencies with the climate change narrative.

Table 5-1
Do US Records for State Temperature Extremes Support a Claim that Climate is Warming Significantly?
Number of US States Having…

All-time *High* set in later year than all-time *Low*	24	48%
All-time *Low* set in later year than all-time *High*	24	48%
All-time *High* & *Low* set in same year	2	4%
All-time *High* set on or after 1950	18	36%
All-time *Low* set on or after 1950	24	48%

The evidence summarized in Table 5-1 answers the five questions posed above: No, no, no, yes, and emphatically no!

Figure 5-13 graphs state extreme temperature records by decade, beginning with the decade of the 1880s.

Based on the number of states (23) having their all-time record high temperature recorded in the 1930s, that decade is certified as, by far, the hottest decade on record. Ironically, the most states setting their all-time record low temperature (9) also occurred during the 1930s.

By far, the decade with the *combined* record for most extreme records set is the 1930s with thirty-two (nearly one-third of all states' extreme temperature records). The decade with the next highest combined all-time records is the 1980s when nearly twice as many all-time low temperature records than all-time high temperature records were set. Ironically, on the heels of nearly four decades (1944–1976) of modest global cooling (-0.4°C; -0.7°F), global warming alarms were first sounded during the 1980s.

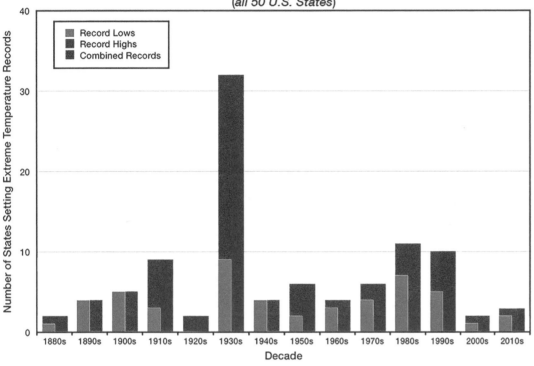

Figure 5-13

Number of States Setting Extreme Temperature Records Per Decade
(all 50 U.S. States)

Source: Wikipedia - U.S. state temperature extremes (https://en.wikipedia.org/wiki/U.S._state_temperature_extremes)

decade is the hottest on record? It doesn't. Because it can't. It simply ignores the evidence.

The evidence of all-time high temperature records by decade indicates that recent climate is actually cooling, an indication confirmed by the USCRN temperature records! A further challenge to theory is the fact that in the 21st-century more states (3) have set their all-time record low temperature than (2) have set their all-time record high temperature! One of those all-time record low temperatures was set in 2019, the most recent all-time state extreme temperature record in the record book.

It is important for the jury to distinguish between individual record high temperatures at a specific location on a particular date and the *all-time record high temperature* for that specific state. With 365 days in most years, there are many, many more opportunities to set a high temperature record in a year than to set an *all-time high temperature record* (there is only one all-time high temperature record for each state). For example, a high temperature record during January at a specific site in a US state might be quite a bit lower than the all-time high temperature recorded for that state. Don't let the prosecution fool you by claiming "hundreds" of record highs for various locations when none match the all-time record high for their respective states.

Curiously, and in stark contrast to what the prosecution's climate change theory expects, the most recent two decades *have the least number* of state all-time record high temperatures of any two contiguous decades in the last hundred years and *the fewest* all-time state temperature extremes of any two contiguous decades since 1900.

How does the prosecution convince the jury that the record temperature extremes in the fifty US states provide any support to either of its claims that climate warming today is unusual or that the current

The prosecution cannot respond to this evidence because it flies in the face of the climate change narrative upon which it has built its case.

With climate warming hysteria coming apart at the seams, will the prosecution turn on a dime and begin predicting catastrophic *global cooling* and blame human activity for the coming global cooling?

Based on the evidence, it would be easier to sell a global cooling story, though a plausible human causation narrative (an apparent requirement) might be a challenge to create.

Ironically, the testimony of state all-time temperature records does a better job supporting global *cooling* than global *warming!*

Summary: Do US Climate Records Support a Claim of Significant Recent Climate Warming?

1. The weighted monthly temperature change shows a *net cooling* of 0.74°F (0.41°C) in the USCRN records from 2005 through 2019. This new evidence is consistent with the evidence presented in Chapter 4 that shows real-world evidence does not portend the feared climate Armageddon claimed by the prosecution. During a period when atmospheric CO_2 grew by 30 ppm, no related temperature change was observed.
2. Except for a small warming shift that coincided with both a strong solar maximum and the 1998 El Niño, the long-term record shows only an historically slight post-LIA natural warming trend, despite nearly two decades of a 21st-century warming hiatus confirmed by satellite, radiosonde, and USCRN records.
3. Satellite and radiosonde measurements are better-controlled, independent, and consistent with each other. The USCRN is the best ground station temperature monitoring network on Earth. All three are considered superior resources for building a record of observed evidence.
4. One-third more US states set their all-time *low temperature more recently* than their all-time high temperature.
5. While nearly half (48%) of all states set their all-time low temperature since 1950, *only 36% of states set their all-time high temperature since 1950.*

6. In terms of the number of states whose current all-time high temperature was set during the 1930s, that decade remains the hottest decade on record. *The 1930s dwarfs all other decades with twenty-three state all-time high temperature records set.* The next closest decades are the 1910s with just six and the 1990s with five (added together, they are still less than half the number of states whose all-time high temperature record was set in the 1930s).
7. Starting with the decade of the 1960s, every decade has either matched or set *more all-time state low temperature records than high temperature records*! A bizarre reality for a period during which "global warming" was so much in the news! A combined twenty-four states set their all-time low temperature records during what the prosecution claims to be a period of dramatic global warming!
8. With 64% of state all-time high temperature records set *prior to* 1950 and 48% of state all-time low temperature records set *on or after* 1950, based on the evidence in these records it would be easier for the jury to conclude that climate is cooling.

Whatever Happened to Global Cooling and That Other Global Warming?

Jurors under the age of fifty may be completely unfamiliar with the great *global cooling* fears of the 1960s and 1970s. Scientists of that era were alarmed that we might be witnessing the end of the Holocene interglacial and the onset of a new ice age.

Since peaking in the 1930s, global temperatures were steadily *trending downward* until the late 1970s.

In the late 1970s, global climate cooling ended as global climate warmed coincident with the approach of a powerful solar grand maximum.

A slight overall net warming consistent with a slow climate warming since the end of the LIA characterizes the overall long-term temperature record from the 19th century to present. Though temporarily boosted by a strong El Niño event during 2015–16, climate change during the 21st century has been marked by a persistent pause in strong warming consistent with the end of the strong solar maximum.

The early baby boomer generation remembers the mid-20th century years of severe winters, cooler summers, and more frequent late winter-early spring snowfalls:

> In the 1970s, major news publications, like *TIME, Newsweek* and *National Geographic*, and significant scientific journals published reports about global cooling. The reports warned that global cooling—a "new ice age"—would devastate food supplies, slash growing seasons across the globe, and would result in widespread famine. Leading scientific organizations, from the National Academy of Sciences to the National Oceanic and Atmospheric Administration's Center for Climatic and Environmental Assessment, warned of unknown dangers related to the cooling phenomenon.

> So serious was the threat that innovative thinkers developed plans to melt Arctic glaciers by covering them in black soot.

Additional plans were offered to divert arctic rivers in the hope that water flows would spread the cooling effect over a broader area. No less a scientific luminary than Carl Sagan, the world-famous physicist who hosted the popular 1980s science series *Cosmos*, suggested the burning and clear-cutting of forests might lead to a new ice age. His suggestion: the intentional release of greenhouse gas emissions to counteract the cooling.[110]

Many scientists expressed fears of the dawn of a new ice age[111] as climate cooling that began circa 1940 continued into its fourth decade. Jurors who are too young to recall the years from the late 1960s to the late 1970s when climate cooling ceased are unlikely to know that, at the time, some scientists expressed serious concerns about *global climate cooling*.

Some attributed climate change to natural cycles, while others blamed human activity and sought to ban the internal combustion engine (sound familiar?). Concerns about dramatic climate cooling and the dawn of a new ice age were in the news nearly as often as the terms *human-caused climate change* and *global warming* are infused into today's news:

> Get a good grip on your long johns, cold weather haters—the worst may be yet to come. That's the long-long-range weather forecast being given out by "climatologists." the people who study very long-term world weather trends... ("Colder Winters Herald Dawn of New Ice Age—Scientists See Ice Age in the Future," *The Washington Post*, January 11, 1970)

The world could be as little as 50 or 60 years away from a disastrous new ice age, a leading atmospheric scientist predicts… ("US Scientist Sees New Ice Age Coming," *The Washington Post*, July 9, 1971)

Ice Age Cometh—Glaciers are growing, warmth-loving animals are moving south, and we all may be getting colder… ("The Ice Age Cometh" in *The Saturday Review*, March 24, 1973)

In the last decade, the Arctic ice and snow cap has expanded 12 per cent, and for the first time in this century, ships making for Iceland ports have been impeded by drifting ice. In England, the average growing season is a week shorter than in 1950, and in the United States, the warm-blooded armadillo is retreating from the Midwest to the South. In Africa, the Sahara desert is creeping southward… ("Is Another Ice Age Due? Arctic Ice Expands In Last Decade" in *Youngstown Vindicator*, March 2, 1975)

There are ominous signs that the Earth's weather patterns have begun to change dramatically and that these changes may portend a drastic decline in food production—with serious political implications for just about every nation on Earth. The drop in food output could begin quite soon, perhaps only 10 years from now. The evidence in support of these predictions has now begun to accumulate so massively that meteorologists are hard-pressed to keep up with it. In England, farmers have seen their growing season decline

by about two weeks since 1950, with a resultant overall loss in grain production estimated at up to 100,000 tons annually… ("The Cooling World" in *Newsweek*, April 28, 1975)

The world's climate is changing… Sooner or later a major cooling of the climate is widely considered inevitable. Hints that it may already have begun are evident. The drop in mean temperatures since 1950 in the Northern Hemisphere has been sufficient, for example, to shorten Britain's growing season for crops by two weeks… ("Scientists Ask Why World Climate Is Changing; Major Cooling May Be Ahead," *The New York Times*, May 21, 1975)

Startled millionaires wintering in their baronial mansions in West Palm Beach, Fla., peered closer last week at the miracle that was falling from the skies and discovered—could it be?—yes, the substance was snow, the first ever reported there. Since mid-November, pedestrians in Dallas, unaccustomed to such hazards, have been slipping on sleet-slicked sidewalks. Meanwhile, a series of blizzards has smothered Buffalo this winter with an astonishing 126.6 in. of snow… ("The Big Freeze," *Time Magazine*, January 31, 1977)

One of the questions that nags at climatologists asks when and how fast a new ice age might descend. A Belgian scientist suggests this could happen sooner and swifter than you might think… ("New Ice Age Almost upon Us?" *The Christian Science Monitor*, November 14, 1979)

Particularly interesting about these reports is that today's obsession with warming is based on the climate of the 1970s being normal and comparing today's polar and mountain glacial ice extent with that of the late 1970s at the end of nearly forty years of persistent cooling.

In the late 1970s climate had reached a multi-decade low point, glaciers were larger, polar ice was at its greatest extent since the LIA, and winters had become more severe than at any other time during the 20th century. Using that relatively frigid point in climate history as a basis for determining whether subsequent climate deviates from "normal" is either ignorant or dishonest.

Both the environmental and political climate of the day are aptly described in a 2008 article by the late Bernard Switalski.

From "Environmentalism's Tainted Roots":[110]

> Beginning in the late 1930s, what passes for the global temperature record took a slight (statistically insignificant) downward trend. Based on that minute dip, in the early 1970s the prophets of doom rang the alarm! *A NEW ICE AGE IS COMING! A NEW ICE AGE IS COMING!"*

> And the culprit? Industrialism(!) and its noxious effluents, dust and smoke, blocking out the sun, threatening to throw the planet into the deep freeze.

> At the time, frights like these appeared in print:

> Reid Bryson, longtime eco-deep-thinker, 1971:

> *"The continued rapid cooling of the earth since World War II is also in accord with increased global air pollution associated with industrialism, urbanization, and exploding population…"*

> Peter Gynne, *Newsweek*, 1975:

> *"There are ominous signs that the earth's weather patterns have begun to change dramatically and that these changes may portend a drastic decline in food production – with serious political implications for just about every nation on earth."*

> Nigel Calder, former editor of the *New Scientist*, 1975:

> *"The facts have emerged, in recent years and months, from research into past ice ages. They imply that the threat of a new ice age must now stand alongside nuclear war as the likely source of wholesale death and misery for mankind"*

> But then, smack in the middle of the campaign to stampede the proletariat into ice age hysteria, the global temperature trend took a slight (statistically insignificant) upward slope.

> Uh-oh!

> Problem?!

No problem.

On a dime, without so much as an, "Excuse my elbow," the prophets of doom spun a one-eighty. By George! It isn't global COOLING that threatens life as we know it! By golly! It's global WARMING!

The jury hears today's prophets of doom utter the same breathless references in reverse, this time fretting about animal migrations and longer growing seasons caused by climate *warming*. Why is it the least bit surprising that animals migrate according to the nature of short-term climate variations? Humans are known to have done the same thing!

When climate warmed during the Medieval Warm Period, Greenland was discovered and settled. When the LIA moderately plunged temperatures, settlers either left or perished as their settlements became buried in accumulating snow and ice.

Animals have been migrating with climate change during millions of years of natural climate change! Evidently, the prosecution believes any climate change *must be* treated as catastrophic and, of course, blamed on humans using fossil fuels.

Global warming doomsaying is rooted in a myopic focus on a chance coincidence of a few years of normal climate warming with strong atmospheric CO_2 growth. Ensuing climate change fears are nurtured by the prosecution's evidence-challenged blind devotion to a theory supported only by poorly performing theory-based climate simulation models!

What could possibly go wrong?

After the natural cycle of multidecadal mid-century cooling ran its course and global temperature resumed its long-term rebound from the Little Ice Age, suddenly rising temperatures were, once again, deemed *unnatural* and likely the result of human activity! The jury might want to ponder just how Earth's climate managed to change *before* humans arrived on the scene since, according to the prosecution, *humans always seem to be responsible* for climate change.

Consider that Earth's climate has undergone a number of warm periods during the current interglacial (Figure 4-23, page 89), most of which peaked at higher temperatures than are observed today. How is it that those warm periods *were not* related to human activity and yet the current modest warm period *"must be"* related to human activity?

On the basis of relatively brief climate warming aided by the 1998 El Niño, like a stroke of lightning, many of those very same scientists who had been warning of a coming ice age bolted to the global warming bandwagon of the late 20th and early 21st centuries.[112]

Since the 1950s, it seems that some scientists have been wringing their hands over the prospect of human-caused climate catastrophe of one extreme or the other.

When global cooling was the rage, it *must be* that atmospheric particles created by fossil fuel combustion were cooling the planet by blocking a portion of solar irradiance.[110] Evidently, in those days, solar irradiance was considered a powerful climate change force. But along came the greenhouse gas climate change theory, and that belief had to be quickly jettisoned.

Now that global warming is the rage, human emissions of CO_2 *must be* responsible for dramatic CO_2 growth, which, in turn, *must be* why the planet is warming.

Got that? Moderately cooling climate is unnatural and moderately warming climate is unnatural and both are caused by fossil fuel combustion from human activity!

There seems to be a common thread here, but it isn't grounded in science. Evidently, some people naively believe climate should just *never* change, so if it *does* change, humans *must be* responsible!

Summary: Whatever Happened to Global Cooling and That Other Global Warming?

1. Multidecadal cooling and warming climate phases experienced over the past 100 years are perfectly normal and well within the expected bounds of natural climate variations during an interglacial of an ice age cycle within an ice epoch of an ice era.
2. The "dust bowl" warming of the 1930s was stronger than either of the recent climate warming spikes, yet that early warming occurred well before atmospheric CO_2 began sharply growing in 1944.
3. Strong warming spikes in recent years are clearly better-related to two strong ENSO events (1998 and 2016). Both events are very evident in the satellite record (Figure 5-1). No observed evidence exists that links such warming to any human activity.

Is the Elephant in the Room Being Ignored?

Every juror has experienced being outdoors on a sunny, summer day when a fair-weather cloud drifts by and briefly blocks the sun. During that brief moment, despite being embedded in a sea of infrared radiation from Earth's surface, a noticeable cooling is felt. This common experience is apparently lost on the prosecutor who maintains that solar variability can safely be ignored by simply assuming atmospheric CO_2 is a much more significant warming force.

There is now considerable research that indicates solar activity (Figure 3-4, page 43) is more likely a driving force for the climate variability experienced during the past two millennia and a much more credible alternative to the prosecution's view that recent climate warming is caused by atmospheric CO_2 growth fueled by the combustion of fossil fuels for energy.

Specifically, the role of solar variability that affects solar irradiance (and insolation) has been downplayed by the prosecution because it does not support the prosecution's climate change narrative.

Sufficient evidence exists to contest the prosecution's view[45] that solar variability is an insignificant climate change factor. From page 4 of *The Solar Evidence*,[113] "NASA Study Finds Increasing Solar Trend That Can Change Climate:"[114]

> Since the late 1970s, the amount of solar radiation the sun emits, during times of quiet sunspot activity, has increased by nearly .05 percent per decade…Historical records of solar activity indicate that solar radiation has been increasing since the late 19th century. If a trend, comparable to the one found in this study, persisted throughout the 20th century, it would have provided a significant component of the global warming the Intergovernmental Panel on Climate Change reports to have occurred over the past 100 years.

From page 5 of *The Solar Evidence*:[113]

> Although the sunspot cycle is approximately 11 years it varies and has generally been getting shorter over the last century. The following figure [Figure 5-14, modified for clarity] shows the "Variations in the air temperature over land in the Northern Hemisphere ([blue] line) closely fit changes in the length of the sunspot cycle ([red] line). Shorter sunspot cycles are associated with increased temperatures and more intense solar activity. This suggests that solar activity is at least partly responsible for the rise in global temperatures over the last century. (Professor Kenneth R. Lang, Tufts University [http://ase.tufts.edu/cosmos/view_picture.asp?id=116])

Note that Professor Lang is not suggesting that sunspot cycle length is the *only* determinant of multidecadal climate variability, but he does observe that the correlation with climate change during the 50 years ranging from 1910 to 1960 is vastly better for sunspot cycle length than for atmospheric CO_2 change (Figures 5-14, 5-15, and 5-16). Note: NH = Northern Hemisphere.

There are a number of factors that contribute to climate variability over multiple decades. Climate and climate change are complex mechanisms in nature. The failure of prosecution scientists to take seriously the full impact of those natural mechanisms is no excuse for them to simply assume that one *alleged* force, changing atmospheric CO_2, is both a significant and dominant force driving contemporary climate change when there is no evidentiary basis for such a claim.

Yet that is precisely what the prosecution has done despite the implications of Figure 5-14[113] that clearly shows CO_2 does not warrant the assumption that it is primarily responsible for recent observed climate variability. Figures 5-14 through 5-16 plot different types of data on different scales and are only shown to illustrate the shape of the curves and the degree of their apparent data correlation. Note in Figures 5-14, 5-15, and 5-16 clear evidence of that pesky theory-inconsistent cooling, circa 1940–1970s, that has since been a fortuitous target of homogenization in an attempt to erase all traces of cooling from the record!

The prosecution response to Professor Lang's chart is to discredit it by claiming the prosecution has some qualms about data sourcing, and according to the prosecution, post-2000 sunspot cycle length correlation with temperature was not as good as the pre-2000 correlation.

Note, however, that in Figure 5-14, the period 1960-1970 is also not well correlated with sunspot cycle length, but that isn't sufficient to discredit the overall good correlation between sunspot cycle length and Northern Hemisphere temperature anomalies. The prosecution's defensive reaction to this evidence is a fine example of grasping at straws while ignoring the broader implications of the evidence.

To those prosecution excuses, the jury should note that CO_2 growth is *much less correlated* with temperature throughout the range of years shown. Furthermore, Professor Lang hasn't claimed that sunspot cycle length is the *only* criterion upon which climate change might be linked (note the correlation with total solar irradiance, TSI, in Figure 5-16).

As noted earlier, Figures 5-14 through 5-16 show that for a period of 120 years, both sunspot cycle length and TSI have far superior correlation with Northern Hemisphere and Arctic temperature anomalies than does atmospheric CO_2. A reasonable juror might wonder why the prosecution refuses to acknowledge its theory does a poorer job of matching reality than does solar activity.

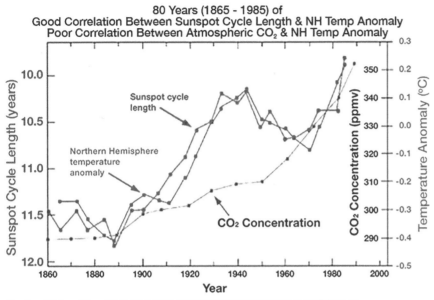

Figure 5-14
80 Years (1865 - 1985) of
Good Correlation Between Sunspot Cycle Length & NH Temp Anomaly
Poor Correlation Between Atmospheric CO_2 & NH Temp Anomaly

Source: Professor Kenneth R. Lang, Tufts University (modified from The Solar Evidence)

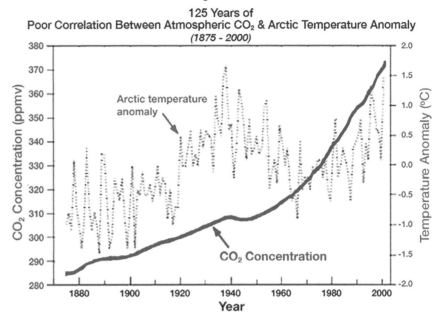

Figure 5-15
125 Years of
Poor Correlation Between Atmospheric CO_2 & Arctic Temperature Anomaly
(1875 - 2000)

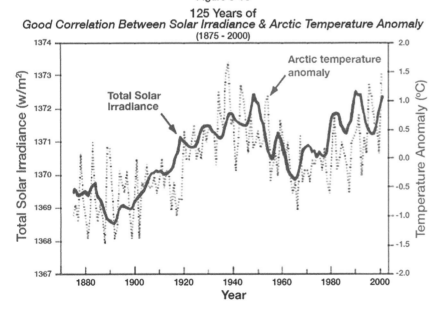

Figure 5-16
125 Years of
Good Correlation Between Solar Irradiance & Arctic Temperature Anomaly
(1875 - 2000)

These observations confirm the findings in Chapter 4 where the lack of any correlation between atmospheric CO_2 growth and global average temperature change is made abundantly clear, compelling evidence for the jury that the relationship between changing atmospheric CO_2 and changing temperature is equivalent to pure chance (the flip of a coin).

Reasonable jurors will conclude from this evidence that there is apparently much more to climate change than atmospheric CO_2 growth (if, indeed, atmospheric CO_2 plays *any* meaningful role at all). In 1985, scientists might have been inclined to believe that climate change is very much linked to sunspot cycle length. Today, the prosecution is more likely to treat such beliefs as heresy against climate change dogma.

Figures 5-15 and 5-16 are based on a 2005 study[115] that looked at the impact of solar irradiance on Arctic temperatures. That study noted solar irradiance in the Arctic (Figure 5-16) is much more strongly correlated with Arctic temperatures than atmospheric CO_2 (Figure 5-15). Based on the evidence of Figures 5-14 and 5-16, it is clear that sunspot cycle length is much more highly correlated with solar irradiance and has a profound impact on Earth's surface temperature.

In view of the 1900–1970 records for CO_2 and temperature shown in Figures 5-14 and 5-15, exactly what motivated prosecution scientists, without seriously investigating any other possible factors, to merely *assume* atmospheric CO_2 growth was responsible for 20th-century climate change?

Why not consider total solar irradiance (TSI)? Was the fortuitous coincident growth of atmospheric CO_2 and Arctic temperature over the limited 1967 to 2000 timeframe sufficient for the prosecution to jump to the enormous conclusion that CO_2 growth was (with 99.9999% certainty) the cause of temperature change? In view of all the evidence, that claim is ludicrous.

To ease comparison, Figures 5-14 through 5-16 were plotted on the same time scale. Which climate force appears to be a better match for temperatures anomalies, atmospheric CO_2 (Figure 5-15, green) or TSI (Figure 5-16, red)? Which would the jury investigate as a more likely source of climate change if jurors had this knowledge in 1985: (1) atmospheric CO_2 or (2) TSI? Note again the evidence of cooling circa 1940–1970s that homogenization of the global temperature record has attempted to eradicate!

What is very apparent is the lack of any meaningful correlation between changing atmospheric CO_2 and Arctic temperatures. Yet there is a good correlation (as might be anticipated) between solar irradiance and Arctic temperatures. This strongly suggests the relationship between sunspot cycle length and solar irradiance is worth exploring.

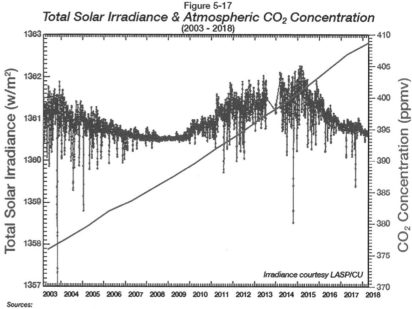

Figure 5-17
Total Solar Irradiance & Atmospheric CO_2 Concentration
(2003 - 2018)

Irradiance courtesy LASP/CU

Sources:
(a) http://cdiac.ornl.gov/ftp/ndp030/global.1751_2014.ems
(b) https://raw.githubusercontent.com/datasets/co2-ppm/master/data/co2-annmean-mlo.csv
(c) SORCE TSI for: February 25, 2003 to April 11, 2018 (Version 17)

An independent atmospheric CO_2 growth comparison using more recent data from Mauna Loa Observatory and the Laboratory for Atmospheric and Space Physics at the University of Colorado (Boulder), compares atmospheric CO_2 growth (in green) with changing total solar irradiance[114] (shown in red) for the years 2003 through February 2018 (Figure 5-17).

These records clearly show that atmospheric CO_2 and TSI during the period from 2003 to early 2018 are not correlated. It is entirely likely that elevated TSI would indirectly raise atmospheric CO_2 by warming Earth's oceans, that, in turn, will release relatively more CO_2 than oceans absorb. On the other hand, warming Earth's landmass would tend to extend growing seasons and thereby absorb relatively more CO_2 from the atmosphere. There is no simple answer to the question, "to what degree does natural climate change affect atmospheric CO_2?"

It is quite possible that atmospheric CO_2 is much more likely *a response* to long term temperature change than the *cause* of it.

Don't Ignore the Clouds

It has been established by experimentation and observation that solar activity has a direct effect on the formation of clouds in Earth's atmosphere. From the article, "Solar Activity Has a Direct Impact on Earth's Cloud Cover," that states:[117]

> A team of scientists from the National Space Institute at the Technical University of Denmark (DTU Space) and the Racah Institute of Physics at the Hebrew University of Jerusalem has linked large solar eruptions to changes in Earth's cloud cover in a study based on over 25 years of satellite observations.
>
> The solar eruptions are known to shield Earth's atmosphere from cosmic rays. However the new study, published in *Journal of Geophysical Research: Space Physics*, shows that the global cloud cover is simultaneously reduced, supporting the idea that cosmic rays are important for cloud formation. The eruptions cause a reduction in cloud fraction of about 2 percent corresponding to roughly a billion tons of liquid water disappearing from the atmosphere.
>
> When the large solar eruptions blow away the galactic cosmic rays before they reach Earth, they cause a reduction in atmospheric ions of up to about 20 to 30 percent over the course of a week. So if ions affect cloud formation, it should be possible to observe a decrease in cloud cover during events when the Sun blows away cosmic rays, and this is precisely what is done in this study.

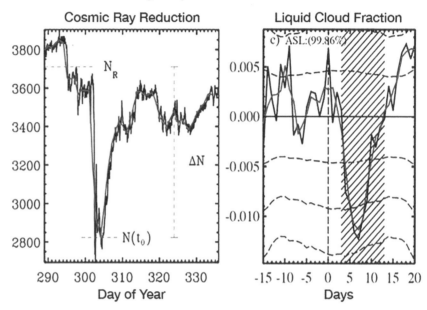

Figure 5-18

Cosmic Ray Impact on Cloud Formation

On the left is shown the cosmic ray reduction (Forbush decrease) following a large solar event. To the right the average response in liquid cloud cover following the five largest Forbush decreases since year 2000 is shown. This shows a clear signal in the clouds -- reduced cloud cover -- follwing the forbush decrease.
(Illustration is based on figures from the new scientific paper)

Based on chart created by: Technical University of Denmark

Figure 5-18 is from the cited article and shows the relationship between cloud formation and cosmic rays. Satellite images of Earth show upper surfaces of clouds appearing bright white, indicating a very efficient reflection of solar radiation.

Note the effect of a passing cloud on a warm sunny day; temperature drops noticeably when a passing cloud blocks sunlight. This is a clear example of the well-established science that, on balance, increased cloud cover leads to cooler temperatures and decreased cloud cover leads to warmer temperatures.

A persistent *decrease* in cloud cover over decades will produce some degree of climate warming. A persistent *increase* in cloud cover over decades will produce some degree of climate cooling. These are far better-established relationships to short-term climate variations than the chance relationship between changing atmospheric CO_2 and climate that the prosecution has used to formulate its dubious GHG climate change theory.

Recall that the Little Ice Age was spawned and sustained by the most notable and prolonged period of "quiet sun" (no sunspots) during the past several millennia (Figure 3-4, page 43). How could the prosecution simply ignore these facts, study conclusions, and their implications?

With clear evidence that other climate change forces are at work, why does the prosecution so desperately cling to a climate change theory, particularly when it is clearly evident that its theory is so at odds with:

- Recorded observations since the late 19th century,
- Ice core evidence during the 10,700 years of the Holocene interglacial, and
- The geologic evidence going back half a billion years?

The evidence *strongly* indicates the most likely source of climate variability is solar variability! Should that really surprise anyone? Is it the least bit shocking that the prosecution has maintained from the beginning that solar variability is far too insignificant to have any serious impact on climate? On what compelling evidence does the prosecution base its dubious claim?

That's right, according to the prosecution, the source of heat (our Sun) can be *safely ignored* as a factor in climate change despite compelling evidence of its strong linkage.

Ironically, there is *no evidence whatsoever* in the observed records to support the prosecution's alleged link between changing atmospheric CO_2 and either temperature or climate change. Indeed, the evidence convincingly testifies that the prosecution's GHG climate change theory is invalid.

Whenever the prosecution is challenged for its allegation that "it may be safely assumed" atmospheric CO_2 growth is responsible for recent climate warming, the prosecution is always ready to pounce with a new "study says" report or an expedient assumption to "settle" the issue:

> Atmospheric CO_2 concentrations had not changed appreciably over the preceding 850 years (IPCC; The Scientific Basis) *so it may be safely assumed* that they would not have changed appreciably in the 150 years from 1850 to 2000 in the absence of human intervention."[118] [*emphasis* added]

Has it never occurred to the prosecution that current "atmospheric CO_2 concentrations" might be entirely irrelevant to climate change? Which would make how much CO_2 concentrations change completely irrelevant to climate change.

The prosecution engages in self-serving speculation by safely ignoring the inconvenient observed relationship between solar variability and climate change that spans the past 1,000 years while "safely assuming" an expedient relationship between atmospheric CO_2 growth and climate change for which there is absolutely no evidence in the records!

The prosecution's brazen assumption-based statement is a stunning revelation that its climate change theory ranks right up there with the tooth fairy as a provable assertion. There is *at least* one elephant in the room, and it is solar variability.

Realizing that acknowledging the role of solar variability would destroy its case, the prosecution is purposefully blind to the existence of the solar elephant.

Summary: Is the Elephant in the Room Being Ignored?

1. Despite solid evidence linking solar grand maxima and minima with corresponding notable climate warming and cooling, the prosecution's climate change narrative has consistently downplayed any significant role for solar variability in observed climate change.
2. There is significant evidence that climate variability is much more strongly correlated with solar variability than atmospheric CO_2.
3. Solar variability directly affects cloud formation that directly impacts solar heating on Earth's surface. Less cloud cover leads to warmer global temperatures; greater cloud cover leads to cooler global temperatures.
4. Persistently strong solar maxima and minima have been linked to warm and cold periods lasting for decades to hundreds of years (Figure 3-4, page 43).
5. While solar variability may not be a perfect match or explain 100% of climate change, it provides a far superior match to climate change than changing atmospheric CO_2 that exhibits only a chance relationship with climate change.

When Computer Models Get It Wrong, Question the Theory Modeled

A particularly egregious example of an apparent prosecutorial effort to deliberately mislead the jury was discovered by Dr. John R. Christy (Professor of Atmospheric Science, Alabama State Climatologist, U. of Alabama, Huntsville) and reported to the House Committee on Science, Space, and Technology in 2017.[119]

Dr. Christy reports he played a minor role as a reviewer during preparation of the (prosecution's) IPCC AR5 report[176]. As such, he made the startling revelation that the IPCC had inadvertently shown that its climate models were in better agreement with real-world observations *only if the models do not include a substantial portion* of atmospheric CO_2.

Because the prosecution assumes that most, if not all, of the recent growth of atmospheric CO_2 is from the atmospheric accumulation of anthropogenic CO_2 (predominantly fossil fuel CO_2), the prosecution ran its models both with and without the extra CO_2 which they alleged to be that portion of atmospheric CO_2 growth attributable to accumulating fossil fuel emissions (i.e., the claimed "human-induced" portion of CO_2 growth from modern civilization's use of fossil fuels).

Christy discovered that the model-projected temperatures agreed very well with observations only when the models *did not include* the full amount of atmospheric CO_2 (when the prosecution's estimate of accumulating fossil fuel CO_2 was not included with the amount of atmospheric CO_2), Figure 5-19. When the actual full atmospheric CO_2 is included (Figure 5-20), the models greatly *overestimate* atmospheric temperature! In other words, the models run "too hot" when the actual atmospheric CO_2 is used; whereas the models only match reality when CO_2 is *substantially lowered below its actual value*.

Figure 5-19

Simplified Chart: Models w/o ACO₂ are Consistent with Observations
Models w/o ACO₂ in blue; Observations in gray

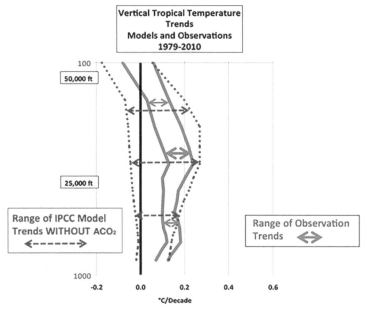

Based on Christy's Figure 5: Simplification of IPCC AR5 shown in (Christy's Figure 4.) above. The blue dotted lines represent the range for the IPCC models, the gray lines are the range of observations. The blue dotted lines are model runs without IPCC's estimate of contribution from ACO₂. Note that non-ACO₂ model runs (blue) overlap the observations almost completely.

Based on: Figure 5 on page 9 of March 29, 2017 Testimony of John R. Christy to U.S. House Committee

Figure 5-20

Simplified Chart: Models with ACO₂ Conflict with Observations
Models w/o ACO₂ in blue; Model with ACO2 in red; Observations in gray

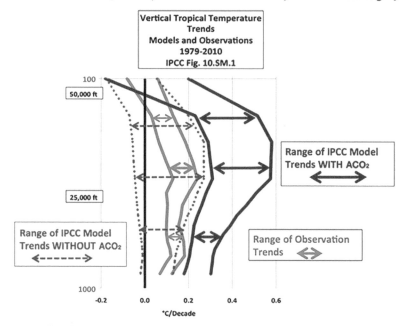

Figure 5. Simplification of IPCC AR5 shown above in Fig. 4. The colored lines represent the range of results for the models and observations. The key point displayed is the lack of overlap between the ACO₂ model results (red) and the observations (gray). The non-ACO₂ model runs (blue) overlap the observations almost completely.

Original source: Figure 5, page 9 of March 29, 2017 Testimony of John R. Christy to U.S. House Committee on Science, Space & Technology. Christy's "Fig. 4" is shown below as Figure 5-21, the AR5 obscure chart.

If the theory upon which the models are based is valid, then the model runs with the full amount of atmospheric CO_2 should closely match actual observations while those runs with atmospheric CO_2 reduced by the prosecution's estimated contribution from fossil fuel CO_2 should project temperatures that are notably cooler than observed. But that didn't happen.

This is what should be expected when the theory upon which the models are based is invalid and asserts atmospheric CO_2 is a powerful climate change force when, in fact, that simply is not true, as jurors have already observed with real-world evidence that shows there is only a chance relationship between growing atmospheric CO_2 and climate change.

Christy's significant discovery is compelling evidence that greenhouse gas modeling is based on an invalid theory as testified to by the array of theory-based CMIP5 models that *consistently* overestimate climate warming from growing atmospheric CO_2.

It should come as no surprise that models based on invalid theory will produce invalid projections. The exposed inconsistency between projections of theory-based models and real-world evidence constitutes compelling testimony that the underlying climate change theory is invalid.

When their underlying theory is bad, no matter how robust the data driving the models, no matter how well model developers create models that reflect a theorized relationship between growing atmospheric CO_2 and atmospheric temperature warming, model projections will be no better than the validity of the theory upon which they are based.

The analysis Christy presented on pages 7–10 of his March 29, 2017, testimony[119] to the US House Committee on Science, Space, and Technology clearly demonstrates "the consensus of the models fails the test to match the real-world observations by a significant margin." This conclusion is a key finding that provides jurors with yet additional compelling testimony that the IPCC-prosecution theory is *deeply flawed* and fails scrutiny by *the scientific method.*

As every good scientist should, Christy used *the scientific method* to compare model results with real-world observations.

As jurors learned from the records examined in Chapter 4, the prosecution's theory-based claims are inconsistent with real-world observations. Christy discovered theory-based models also fail to agree with real-world observations.

Both failures exist for the same reason. The prosecution's theory that changing atmospheric greenhouse gases are a strong climate change force is simply not a valid theory.

Burying Exculpatory Evidence!

When Christy insisted that the significant discovery he made be presented in the main text of the AR5 document, "the government-appointed lead authors decided against it." Why? The jury should think long and hard about why the prosecution *buried* such clear evidence that its climate change theory is invalid! Would an honest prosecutor bury such exculpatory evidence?

Christy learned the IPCC (prosecution) had:

> ...inadvertently provided information that supports this conclusion by (a) showing that the tropical trends of climate models with extra greenhouse gases failed to match actual trends and (b) showing that climate models without extra greenhouse gases agreed with actual trends. A report of which I was a co-author demonstrates that a statistical model that uses only natural influences on the climate also explains the variations and trends since 1979 without the need of extra greenhouse gases. While such a model (or any climate model) cannot "prove" the causes of variations, the fact that its result is not rejected by the scientific method indicates it should be considered when trying to understand why the climate does what it does.

Instead of giving significant exposure to this dramatic discovery, what the prosecution actually did is perhaps even more appalling than its dogged refusal to admit its climate change theory is invalid.

Rather than reveal Christy's discovery, the prosecution *obscured the discovery* by hiding the charts that exposed this substantial failure "in the Supplementary Material where little attention would be paid…" As if to punctuate the prosecution's arrogant determination to bury this important finding, the charts concocted to display these model runs were fashioned "in such a way as to make it difficult to understand and interpret."

Figure 5-21 is the chart the IPCC designed to obscure their own model runs that discredit their theory—vertical direction is increasing altitude, left to right in each panel is increasing temperature. Compare Figure 5-21's obscure meaning with the clear message conveyed by Christy's charts in Figures 5-19 and 5-20.

It is crystal clear that the prosecution's presentation of this material was deliberately obtuse, difficult to analyze, and cleverly buried in the bowels of working group Supplementary Material to the IPCC AR5 report. Why was that done?

If human-induced climate change were really significant and so obvious, then why would the prosecution use subterfuge to continually try to *deceive the jury* and then hide the evidence of their deception? The prosecution deliberately buried exculpatory evidence to atmospheric CO_2 produced by the prosecution's own climate simulation models that exposed the prosecution's climate change theory as fantasy.

A more traditional presentation of CMIP5 models' overestimating greenhouse gas temperatures is illustrated by Figure 5-7 on page 101 where the solid red line is (again) the average projection from prosecution's climate simulation models.

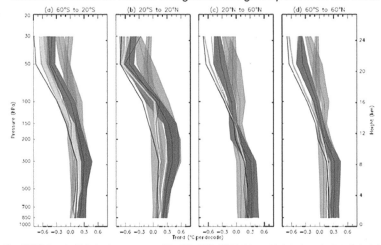

Figure 5-21
IPCC Chart in Supplemental Materials
Obscure chart shows models with greenhouse gases performed the worst

Figure 10.SM.1 | Observed and simulated zonal mean temperatures trends from 1979 to 2010 for CMIP5 simulations containing both anthropogenic and natural forcings (red), natural forcings only (blue) and greenhouse gas forcing only (green) where the 5th to 95th percentile ranges of the ensembles are shown. Three radiosonde observations are shown (thick black line: Hadley Centre Atmospheric Temperature data set 2 (HadAT2), thin black line: RAdiosonde OBservation COrrection using REanalyses (RAOBCORE) 1.5, dark grey band: Radiosonde Innovation Composite Homogenization (RICH)-obs 1.5 ensemble and light grey: RICH-τ 1.5 ensemble. (Adapted from Lott et al (2013) but for the more recent period from 1979 to 2010.)

Figure 4. This is Fig. 10.SM.1 of the IPCC AR5 Supplementary Material for Chapter 10. These are trends (1979-2010) for various vertical levels of the atmosphere from (a) observations (gray band – difficult to see), from (b) models without extra GHGs (blue band) and (c) models with extra GHGs and other forcings (red band). The lower portion of the tropical chart (second panel from left) is simplified in Fig. 5 and used for the following discussion.

Source: page 8 of March 29, 2017 Testimony of John R. Christy to U.S. House Committee on Science, Space & Technology

By projecting much higher temperatures than are actually observed, both Figures 5-7 and 5-20 offer jurors clear and compelling evidence the prosecution's models are in serious conflict with real-world observations!

The two gray lines in Figures 5-19 and 5-20 represent the boundaries of real-world observations. Jurors should note that the prosecution's models project within those boundaries (agree with real-world observations) *only* when models use a much lower concentration of greenhouse gases than is measured in the atmosphere; whereas when models use the correct concentration of greenhouse gases they vastly over-project atmospheric temperature. This is clear confirmation that the theory modeled is completely invalid.

What does this sorry episode of prosecutorial misconduct suggest about the prosecution's greenhouse gas climate change theory that is the basis upon which misfiring climate simulation models are developed?

Important notes for jurors:

- When the actual measured value of atmospheric CO_2 is used to drive its climate simulation models, the prosecution's theory-based models vastly *over-project* temperatures.

- When a much lower value for atmospheric CO_2 is used to drive its models, they *accurately project* real-world temperatures!

What result would the jury expect from a computer simulation developed on the basis of an invalid theory that vastly overestimates the climate impact of atmospheric greenhouse gases?

Would jurors expect real-world observations to agree with computer simulations built on a faulty foundation? Of course not.

When the prosecution deliberately deceives the jury by burying this important finding in obscurity, it commits the worst kind of prosecutorial misconduct. Such behavior is inexcusable and suggests dishonesty has become an essential tool of the prosecution's efforts to dupe the jury while promoting its deeply-flawed theory-based assault on fossil fuels.

Recall the discussion of *the scientific method* in which a famous quote from renowned physicist Richard Feynman was paraphrased:

> It doesn't matter how beautiful your theory is, it doesn't matter how smart you are. If it doesn't agree with observation, it's wrong.

More to the point:

> It doesn't matter how beautiful the prosecution's theory is, it doesn't matter how smart prosecution's scientists are, if prosecution's theory fails to agree with observation, its theory is flawed and *must be* rejected as invalid.

Christy further discovered the models overestimated real-world observed temperatures by *85% for global temperatures and 142% for tropical temperatures*. Christy concluded:

> The scientific conclusion here, if one follows the scientific method, is that the average model trend fails to represent the actual trend of the past 38 years by a highly significant amount. As a result, applying the traditional scientific method, one would accept this failure and not promote the model trends as something truthful about the recent past or the future. Rather, the scientist would return to the project and seek to understand why the failure occurred. The most obvious answer is that the models are simply too sensitive to the extra GHGs [greenhouse gases] that are being added to both the model and the real world.

Here *the scientific method* is at work, being used properly by a credible, competent scientist.

Contrast Christy's work with the work of prosecution scientists who shun *the scientific method* in favor of expediency and a stubborn refusal to reconsider their deeply-flawed theory despite the testimony of a mountain of contradictory evidence.

When theories (or theory-based models) fail to agree with real-world records, then it isn't the records that need to be homogenized or otherwise doctored. The models are wrong because the theory driving them is wrong! That same flawed theory is the very foundation for the prosecution's case against atmospheric CO_2 and fossil fuel emissions.

Sooner or later, the prosecution must accept the fact that nothing can salvage computer models that are built to the specifications of an invalid theory. Not even cooking the books by altering historic temperature records can resuscitate a fatally-flawed theory.

Summary: When Computer Models Get It Wrong, Question the Theory Modeled

1. The massive array of CMIP5 computer simulation models based on the prosecution's greenhouse gas climate change theory are shown to be in disagreement with real-world observations.
2. When the prosecution's climate simulation models are run *with a much lower concentration* of atmospheric CO_2 than is measured in the atmosphere, the models project observed temperatures.
3. When models are run *with* the actual measured concentration of atmospheric CO_2, they vastly over-project temperatures (they run "hot").

4. When models disagree with real-world data, then either the models contain errors or the theory upon which they are based is flawed.
5. *The scientific method* requires flawed theory and/or models to be rejected or sufficiently altered until they conform with real-world observations.

Is Modern Warming Really Unprecedented?

Whenever climate change is being discussed, there is often a reference to global warming and unprecedented atmospheric warming. Jurors have already seen that over the past 550 million years, the dominant climate has been *much warmer* than any climate humans have ever experienced.[38]

But what about the ice era of the past 60–65 million years?

An ice era is a very cold deviation from Earth's typical very warm climate, so is current warming really *unprecedented* in the current ice era? Figure 5-22 shows North Pacific temperatures during the past six million years (~10% of the current ice era) including the 2.4 million years of the current (Pleistocene) ice epoch. Current climate is at the extreme right side of the chart.

Barely discernible at the border on the right side of the red curve, the current interglacial (Holocene) temperatures peak just at the horizontal bar ("*Now*"). Note the ice age-interglacial cycles during the Pleistocene ice epoch with several prior interglacial *peak temperatures at or exceeding* current Holocene interglacial temperatures.

The scale of this chart does not allow for detail; however, Figure 4-1 (page 57) is an expanded higher resolution view of the 800,000 years found between the "0" and "1" million years of this chart and the entire Holocene interglacial (Figure 4-23, page 89) is compressed into the barely discernible red line at the right side of the graph.

The peak climate of the current interglacial is roughly equivalent to the coldest climate between 3 and 6 million years ago, with most of the climate prior to 3 million years ago being much warmer than any climate humans have experienced yet still much colder than Earth's typical climate and even colder than the climate during most of the earlier portion of the current ice era that began about 60 million years ago (Figure 3-5, page 46).

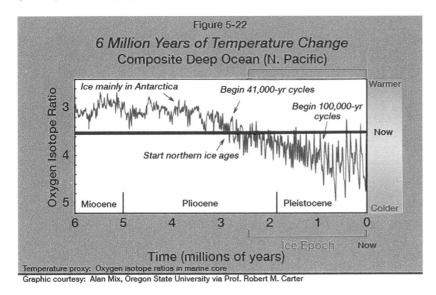

Figure 5-22

6 Million Years of Temperature Change
Composite Deep Ocean (N. Pacific)

Temperature proxy: Oxygen isotope ratios in marine core
Graphic courtesy: Alan Mix, Oregon State University via Prof. Robert M. Carter

The prosecution's claim that current global temperatures represent "unprecedented" warmth is utter nonsense and further evidence of prosecutorial misconduct for deliberately trying to mislead the jury by making no effort to put current climate into a proper perspective.

Summary: Who are the *Real* Climate Deniers? More Defense Testimony

If the prosecution's climate change theory were truly scientific and supported by the testimony of real-world observations, there would be an abundance of real-world evidence to support it.

By *looking out the window* at the real-world evidence, the jury has not seen *any* evidence supportive of the prosecution's theory. To the contrary, the jury has seen ample credible evidence spanning decades, centuries, millennia, hundreds of thousands of years, and hundreds of millions of years, each providing clear evidence of significant contradictions, each of which are sufficient to invalidate the prosecution's theory.

At what point does it make any sense to continue to pretend the prosecution's greenhouse gas climate change theory-based allegations against atmospheric carbon dioxide have any merit? All the prosecution has is an invalidated theory, an array of computer simulation models based on that invalid theory, and a raft of speculative what-if studies commissioned to support a scenario (human-induced global warming caused by fossil fuel emissions) for which no evidence in the records exists!

At this point it is clear the prosecution is not engaging in science; it is engaging in systematic speculation designed to support its charter-driven dogma. On a massive scale, its actions are indistinguishable from fraud.

Among the key elements of testimony presented to the jury in this chapter are:

1. The satellite-based lower atmosphere temperature evidence shows a normal post-LIA slight upward temperature trend between 1979 and 1998. The post-2002 (2002–2019) satellite evidence shows no meaningful trend. Known as the pause, this period is clearly evident in both the satellite and

radiosonde records. The 2005 to 2019 USCRN records not only do not show any warming, they show a slight cooling!

2. Net temperature warming since 1979 represents both a recovery from mid-century cooling and natural warming associated with the late 20th-century solar grand maximum. The chance correspondence of this natural warming with atmospheric CO_2 growth spawned global warming speculation based on the invalidated greenhouse gas climate change theory.

3. There is no evidence of any unusual temperature warming unrelated to (a) late 20th-century solar grand maximum or (b) strong ENSO events or (c) recovery from mid-20th-century pronounced cooling. There is no justification for the prosecution's campaign to demonize either atmospheric CO_2 or any portion of atmospheric CO_2 growth attributable to fossil fuel use.

4. The prosecution's greenhouse gas climate change theory requires the appearance of a strong GHG warming hot spot in the tropical (+/- 30° latitude) mid-troposphere. Yet to this date no credible evidence of this required hot spot has been found. In fact, a "not spot" of slight *cooling* has been observed in the region where greenhouse gas theory demands evidence of strong hot spot warming! If the science requires it, yet it isn't there, what does this suggest about the validity of the prosecution's climate change theory?

5. Prosecution theory requires (and theory-based models project) warming in polar regions will be more pronounced and occur *prior to* theory-based warming in lower latitudes. This polar amplification of greenhouse warming is a robust requirement rooted in the prosecution's greenhouse gas climate change theory. As with the consistently missing GHG hot spot, measurements show a failure of the required polar amplification to develop. Some evidence indicates Antarctic polar regions are actually *cooling*. This is yet another exam-

ple of real-world observations contradicting the prosecution's climate change theory!

6. When historic temperature records contradict the prosecution's climate change theory, the records have been subject to questionable *homogenization* (i.e., adjusting) that magically made many records appear to better support theory! There is no useful scientific purpose served by ongoing data manipulation that conveniently adjusts recorded temperatures to create an *illusion* that the prosecution's theory of a crime is valid. This fraud is accomplished by consistently lowering past temperatures and elevating more recent temperatures. It is unlikely such a coincidence of fortuitous adjustments occurs by chance.

7. When the prosecution breathlessly announces recent years are "the hottest on record" or "one of the ten hottest years ever recorded," it apparently expects the jury will not consider the following:

   a. Since Earth's climate is emerging from the LIA that ended circa 1850, wouldn't a long-term average warming trend be normal? If not, wouldn't Earth still be experiencing the LIA?

   b. Decadal length climate variations are normal and superimposed over the underlying slight warming trend as Earth's climate emerges from the LIA into what may become known as the Modern Warm Period. When a natural short decadal warming cycle is superimposed over the normal long-term warming trend since the end of the LIA, is it the least bit surprising to have times when the hottest year on record is noted?

   c. When temperature records extend only as far back as the latter years of the very cold Little Ice Age, shouldn't recording the hottest year on record now and then be expected and viewed as perfectly normal?

d. Is there any basis for contending recent slight climate warming is anything but typical of Earth's climate emerging from a cold period within an interglacial of an ice age cycle during an ice epoch of an ice era?

e. Is it reasonable to contend that recent perfectly normal climate warming consistent with past warming is somehow different this time and it *must be* caused by the contribution of fossil fuel emissions to atmospheric CO_2?

8. Since individual state all-time high temperature records began being kept, US fifty-state extreme temperature records show no pattern of warming consistent with the prosecution's climate change theory. Since 1950, more states have set their all-time low temperature record than their all-time high temperature record. Does that suggest climate warming or cooling? 64% of states have set their all-time high temperature record *prior to* 1950. Doesn't this suggest recent climate *is not* dramatically warming as the prosecution contends? The 1930s remains, by far, the decade with the record number (23) of states setting their all-time high temperature record, *nearly four times higher* than the next hottest decade, the 1910s, with just six state all-time high temperature records.

9. Prosecution agents predicted snowfall and snow sledding would be just a distant memory by the year 2013. Yet the winter of 2014–15 was proclaimed by The Weather Channel as "the all-time snowiest" in Boston weather history with "four of its top five snowiest seasons" coming during the previous 21 years. It's not likely those "no more sledding" predictions were based on there being *too much snow* for sledding!

10. The prosecution virtually ignores the solar connection to climate that is *demonstrably far stronger* than any theorized influence atmospheric CO_2 has on climate change. There is a clear connection between cloud formation and cosmic radiation and between cosmic radiation and variable solar activity. During intense solar activity, *cloud cover is reduced*, enhancing warming on Earth. During periods of relatively quiet solar activity, *cloud cover is enhanced*. Together with reduced solar activity, these work to cool climate.

11. Annual changes in outgoing IR do not match annual changes in lower atmospheric temperature anomalies over the period 1979 to 2002. If greenhouse gases were a strong climate change force, then the impact of *changes to outgoing IR should be clearly manifested in the record of lower atmospheric temperature anomalies*. The actual relationship is tenuous at best; further evidence the greenhouse gas climate change theory is inconsistent with real-world evidence.

12. The prosecution's theory-based CMIP5 climate simulation models provide compelling evidence their underlying theory is invalid. When models are run with a much-reduced concentration of atmospheric CO_2 they *match* current climate. Yet when the models are run with the actual measured concentration of atmospheric CO_2, they vastly over-project temperatures, disagreeing markedly with much cooler observed measurements. Since the CMIP5 models are based on the prosecution's theory, the reasonable conclusion is that the theory's basis (that atmospheric CO_2 is a strong climate force) is incorrect and therefore theory and theory-based models will not reflect reality. Put another way, the theory is invalid, and anything derived from theory-based analyses is fruit of the poison tree (i.e., worthless). This evidence alone, together with other compelling supporting evidence from Chapters 4 and 5, is sufficient to declare the prosecution's theory invalidated by *the scientific method* for its failure to agree with observed evidence in the real world.

CHAPTER 6

Summation of Carbon Dioxide's Defense

Occam's razor: A principle that holds that the best explanation is most likely the least complicated requiring the fewest assumptions.

Jurors who have a clear understanding of the evidence they have carefully examined are best prepared to make a true verdict in this trial of carbon dioxide and fossil fuels in the court of public opinion. To that end, the defense offered jurors a heavy dose of evidence observed in nature so jurors won't blindly believe the prosecution's allegations. Blind belief is never a good substitute for truth.

The prosecution's case against atmospheric carbon dioxide is rooted in a theory that claims:

- *If global average atmospheric CO_2 increases, global average surface temperature will warm.*
- *If global average atmospheric CO_2 decreases, global average surface temperature will cool.*

This theory can be simplified to:

- *If global average atmospheric CO_2 changes, then global average surface temperature will change accordingly.*

A practical example of the prosecution's theory:

The prosecution routinely asserts that fossil fuel use causes CO_2 emissions to accumulate in the atmosphere ("*CO_2 increases*") causing global warming ("*temperature will warm*").

If the prosecution's theory were valid, then evidence supporting its claimed relationship between growing atmospheric CO_2 and climate warming would be readily observed in Earth's global records for atmospheric CO_2 and temperature.

The prosecution's key theory-based charge is:

> Recent climate warming is substantially caused by atmospheric CO_2 growth attributable to fossil fuel emissions.

Is this true? While it is true that climate has warmed slightly and that atmospheric CO_2 has grown considerably, in the absence of compelling real-world evidence supporting this theorized relationship, it is unscientific to either assert any causal relationship between either CO_2 growth and climate warming or any claim that CO_2 growth is substantially caused by fossil fuel emissions.

If a theory says condition A leads to response B (*if A, then B*) and then B is found to be independent of A (*B is uncorrelated with A*), then A cannot lead to B and that theory *is invalid.*

If the observed relationship between temperature (climate) change and atmospheric CO_2 growth is contrary to the prosecution's theorized relationship, then that theory must be invalid. Every valid theory *must consistently be* supported by observed evidence.

In correspondence with a senior advocate for the prosecution's climate change theory, a single piece of evidence was offered that describes a compelling contradiction of observed evidence to the prosecution's climate change theory (see Defense Exhibit D, "1880–2018 NOAA, NCDC, MLO Evidence").

Acknowledging the evidence contradicts climate change theory, nevertheless the theory advocate offered an explanation that claimed theorized warming isn't observed in the records because 93% of the warming is being stored in the oceans, 30% of which resides in deep ocean waters.

According to that explanation, only 7% of greenhouse climate change warming is being observed in the atmosphere. Is that consistent with frequently heard dire warnings and headlines? Is it consistent with the many diagrams purporting to show how "backradiation" is warming near-surface *air* temperatures? Is it consistent with stories that warn of "catastrophic" global climate warming that will destroy life as we know it?

Does this claim make any sense at all? Or is it really just an attempt to rationalize why climate isn't behaving according to the prosecution's theory?

Land surfaces warm much more quickly than ocean waters. As post-LIA climate slowly warms, ocean waters are slow to respond. Fill a glass bowl with cold water. Place a heat lamp over the water-filled bowl. Open both hands with palms facing up. Submerge one hand in the water-filled bowl and put the other hand under the lamp at about the same distance from the lamp as the bowl. Note which hand warms more quickly. The same process brings land surfaces to a higher temperature more quickly than ocean waters. So how is it that we are expected to believe oceans are absorbing greenhouse gas warming more quickly and it's all because we use fossil fuels? Really?

With lands covering 29% of the planet's surface, how is it that only 7% of theorized warming is being observed on land?

There are a number of other significant problems associated with claiming the heat "is in the oceans" (presumably just waiting to jump out at us). How is the provenance of ocean heat being certified? Or is its source just being speculated? What portion of ocean warming might be caused by cycles of increased geothermal heating? What portion might be due to solar variability? What portion might be a consequence of ENSO and North Atlantic Oscillation (NAO) cycles?

How do atmospheric molecules heat ocean molecules without affecting atmospheric temperatures?

On a par with "the dog ate my homework," this claim has the appearance of yet another convenient excuse for why the prosecution's theory is contradicted by the evidence in nature. Such excuses also attend the missing theory-required mid-tropospheric tropical warming and the lack of polar amplification (both described in Chapter 5).

Provenance cannot be certified, so (once again) theory advocates *speculate* that ocean warming *must be* caused by changing CO_2, but not to the extent, evidently, that such changes would be noticed in the surface temperature records!

Apparently, jurors are also expected to accept the notion that warmer ocean surfaces do not create warmer air over ocean surfaces. Evidently, jurors are expected to accept without question every explanation for why theorized warming isn't evident in the records.

Since 29% of Earth's surface is covered by land, wouldn't it be more reasonable for jurors to expect that at least 29% of theory-predicted atmospheric warming would be felt on land?

There is also the question of which relationship dominates, climate change causing ocean warming or ocean warming causing climate change.

In Chapters 3–5, the clear and compelling testimony of real-world evidence has conclusively shown that over the same increments of time, *changing* atmospheric CO_2 is *uncorrelated* with both temperature and climate *change*.

The lack of any correlated relationship between changing atmospheric CO_2 and climate change has been demonstrated for the following:

- 25 million year increments of the 550-million-year geologic climate record,
- 100+ year increments of the GISP2 ice core climate record,
- 30-year increments of time in the US government's contemporary temperature records since 1880.
- Annual (year-to-year) increments of time in the U.S. government's contemporary temperature records since 1880.

Based on the testimony of these records, the stunning lack of any correlation between changing atmospheric CO_2 and changing temperature (or climate) over *any* meaningful increment of time during the past half-billion years means the *alleged* causal relationship between changing atmospheric CO_2 and changing temperature (or climate) cannot possibly exist! This constitutes sufficient unequivocal evidence for *the scientific method* to declare the prosecution's underlying climate change theory *completely invalid*. To be consistent with the conclusions of *the scientific method*, the jury *must declare* both atmospheric CO_2 and civilization's use of fossil fuels to be entirely innocent of any discernible impact on global climate.

The defense has found a compelling *preponderance* of evidence over a wide range of timeframes during which atmospheric CO_2 and climate are entirely uncorrelated. On the other hand, the prosecution has not produced any consistent evidence over any meaningful timeframe during which changing atmospheric CO_2 is positively correlated with changing global average temperature. Under these circumstances, no causal relationship is possible between atmospheric CO_2 change and climate change. The evidence in nature unequivocally invalidates the prosecution's theory and refutes its allegations against atmospheric CO_2 and fossil fuels.

In the single record for which a cursory view would suggest atmospheric CO_2 and climate are correlated (Antarctica's Vostok Station 800,000-year ice core record, Figure 4-1, page 57), higher resolution views reject any correlation that supports atmospheric CO_2 change as a driver of climate change since CO_2 change is observed to *lag behind* climate change by an average of *800-years!* While Figure 4-1's scale leaves a misleading first impression, the precedence of climate change to atmospheric CO_2 change in this record is a powerful contradiction to the validity of the prosecution's greenhouse gas climate change theory.

While the behavior of atmospheric CO_2 is perfectly plausible as *a consequence of* changing climate, there is no plausible explanation for why atmospheric CO_2 would change *as a cause of* climate change.

By simply *looking out the window* at the recorded evidence, the jury has seen compelling evidence that unequivocally testifies against any relationship beyond pure chance between atmospheric CO_2 change and temperature (or climate) change.

While the prosecution's theory is complex (and requires a good dose of speculative assumptions about the atmospheric behavior of CO_2), all that complexity and speculation mean absolutely nothing if the theory being offered defies real-world evidence. That theory-defying evidence comes in the form of measured contemporary records, ice core records, and geologic records that offer a great body of evidence upon which the validity of the prosecution's climate change theory has been evaluated in accordance with *the scientific method*.

The evidence compels *the scientific method* to conclude the prosecution's theory is wholly invalid. Since that invalidated theory is the foundation for allegations that both defendants (atmospheric CO_2 growth and fossil fuel use) are responsible for global climate change, then the jury has no recourse but to find both atmospheric CO_2 in

general and fossil fuel emissions in particular completely exonerated from having any detectable impact on contemporary global climate change. There simply is no legitimate basis for the jury arriving at any other verdict.

Recall the key questions posed in Chapter 2:

1. Is recent observed climate change either unusual or significant?
2. Does any evidence exist to confirm a relationship that changing atmospheric CO_2 causes corresponding climate change?
3. Is recent observed atmospheric CO_2 growth really caused by modern civilization's use of fossil fuels?

Jurors can now confidently answer these questions based on the considerable evidence presented in Chapters 3–5 that offers compelling testimony that:

1. Recent observed climate change is neither unusual nor significant.
2. There is no causal relationship between atmospheric CO_2 growth and global warming. The observed relationship is indistinguishable from pure chance. There is no correlation between atmospheric CO_2 change (growth) and climate change (warming). No correlation guarantees no causation.
3. Since atmospheric CO_2 growth has no impact on global climate, the source of recent atmospheric CO_2 growth is entirely irrelevant.

The real-world evidence in every record examined unequivocally contradicts the prosecution's claimed relationship between changing atmospheric CO_2 and global average surface temperature change (or climate change).

The prosecution has an obligation to do more than simply cite isolated examples of short-term chance correspondences between changing atmospheric CO_2 and changing climate in a desperate effort to assert its desired conclusion. It must offer bona fide explanations for why the great preponderance of evidence relentlessly testifies *against* its climate change theory over time periods ranging from decades to tens of millions of years. But it does not.

In addition to offering compelling evidence that fully exonerates atmospheric CO_2 from having any discernible impact on climate change, the jury has seen evidence that exposes the prosecution's climate change theory as invalid speculation. Among the evidence:

1. There are two theory-required manifestations of greenhouse gas climate change theory: (a) the existence of a strong tropical mid-tropospheric hot spot, and (b) polar amplification of greenhouse gas climate warming. Neither of these have been observed despite their existence being an absolute requirement of the prosecution's climate change theory.

2. Climate change theory ignores the key roles played by both variable solar activity (changing TSI) and cosmic radiation's impact on cloud formation and their influences on climate.

3. Claims of hottest year on record ignore the fact that most records date back only to the 1880s. Warming since the end of the LIA should, on occasion, naturally be "the hottest on record." Such occurrences are to be expected, not deceptively used to support a false climate change narrative.

4. The US all-time high and low temperatures records for each state do not indicate any significant climate warming coincident with the rapid increase in atmospheric CO_2 since the mid 20th-century. The decade of the 1930s holds far more all-time state high temperature records than any other decade since 1880.

5. The IPCC's CMIP5 global climate simulation models are built upon the IPCC's greenhouse gas climate change theory. These models provide strong testimony to their underlying theory being invalid. The CMIP5 models habitually overestimate projected warming from atmospheric carbon dioxide. When atmospheric CO_2 *is significantly reduced in the simulations*, the models match reality, demonstrating the theory underlying the models is deeply flawed because it consistently and significantly overestimates surface warming from greenhouse gases, CO_2 in particular.

6. Unremarkable recent climate warming is not unprecedented. Early Holocene climate was warmer, and more recent warm periods (Medieval, Roman, Minoan, etc.) are characteristic of the Holocene interglacial, all of which occurred long before modern humans used the first ounce of fossil fuel.

7. Primarily basing its claims on evidence since 1980, the prosecution conveniently ignores that climate had been slowly cooling for decades until the end of the 1970s. Basing temperature comparisons on the conditions that existed at the end of multiple decades of cooling will always overemphasize the appearance of warming.

8. Claims of unusual warming ignore the fact that Earth's climate is in an unusually cold regime and has been for the past 60 million years. The current interglacial is likely to be followed by another ice age cycle within an ice epoch of an ice era that began 60 million years ago. Earth's typical climate is about 13°F warmer than any climate modern humans have known.

Defense testimony has given the jury incontrovertible compelling evidence that the prosecution's speculative climate change theory, the basis for its human-caused climate change narrative, is unequivocally invalidated by *the scientific method* for the theory's failure to agree with real-world evidence over any meaningful period of time (from contemporary year-to-year change to 25-million-year increments of change in the geologic record).

Consequently, every naive effort (e.g., *Paris Agreement, Green New Deal, emissions restrictions*) intended to alleviate human-caused climate change by reducing atmospheric CO_2 is doomed to failure because (1) no human causation exists and (2) no better than a chance relationship exists between the changing concentration of atmospheric CO_2 and climate change.

No impact on climate change is possible from any effort to control carbon dioxide emissions because the evidence clearly shows that CO_2 emissions have no detectable impact on either temperature or climate change.

The jury has seen ample real-world evidence the prosecution's case is based entirely on a false premise, the theory that changing atmospheric greenhouse gases are a strong *climate change* force. This notion has been invalidated by *the scientific method* for its demonstrated lack of support from real-world evidence.

The testimony given in Chapters 3–5, together with additional evidence in Defense Exhibits A-D provide more than sufficient compelling evidence for the jury to confidently and unequivocally reject *every prosecution allegation* that fossil fuel use, emissions of CO_2, and total atmospheric CO_2 change are strong climate change forces.

After viewing all the evidence, jurors have everything they need to confidently conclude that climate change *is entirely unrelated to atmospheric carbon dioxide growth* regardless of the source of that growth.

No legitimate climate change basis exists to support any public policy to restrict the use of fossil fuel as an abundant and relatively inexpensive source of energy. Regulatory restrictions on CO_2, the gas of life, are costly burdens on the global community that will neither benefit Earth's environment nor any life that inhabits that environment.

While alternate energy sources should be explored for practical reasons (eventually fossil fuels will be exhausted), there is no need to squander public funds on immature alternate energy technology when there is ample motive for private capital to invest in and develop new alternate energy sources at a pace consistent with need and emerging technologies.

Wrong-headed top-down regulatory mandates based on false premises and ignorance are poor substitutes for innovation based on economics, needs assessments, genuine science, and the promise of a commensurate reward for competitively developing the best new solutions to civilization's future sustainable energy requirements.

Changing climate is a fact of life on Earth as it has been for billions of years. The wise course is not found along the path of misguided attempts to alter global climate by over-regulating energy production that, in the end, has absolutely no ability to affect climate change. Better to set a course that anticipates future adaptation to likely natural climate change whose consequences include both coastal retreat *and* coastal inundation.

The jury is encouraged to read the material in Defense Exhibits A through D for additional support of its evidence-based verdict. The jury must not allow its verdict to be influenced by unsupportable prosecution allegations, or poorly-informed political or social pressures.

As this trial draws to an end, each juror should help ensure that others will reach a proper verdict by sharing the knowledge and understanding gained by *looking out the window* at the real-world evidence.

As noted earlier, the specious nature of the prosecution's climate change theory has duped many people, scientists included, into naively accepting the climate change narrative without *looking out the window* at the real-world evidence.

While there may be a natural emotional response to having been duped by the prosecution's persistent pursuit of its climate change narrative, it is better to set any anger aside and just share this material with others to help them understand that real-world evidence invalidates the theory-based claims of the prosecution's climate change narrative. Keep in mind this advice offered in Chapter 2:

> Just as in its past and without any human influence, Earth's future climate will naturally include both bitter cold and steamy heat. At some point, rather than dwell on the false claims of an invalid theory, civilization will do well to prepare for inevitable dramatic *natural* climate change episodes that cannot be prevented.

Truth is not found in blind belief. Jurors have skeptically scrutinized theory by examining the evidence in nature. The jury has followed the path to understanding and truth and is now prepared to render a fair and just verdict.

CHAPTER 7

Defense's Closing Statements to the Jury

By *looking out the window* the jury has seen the clear testimony of *real-world evidence* given by atmospheric carbon dioxide and surface temperature records. These records have *unequivocally refuted* any notion that changing atmospheric carbon dioxide has any discernible impact on contemporary climate change.

This suggests a perfect characterization of the prosecution's dogged pursuit of its climate change theory is the popular observation:

> *It's not what you don't know that gets you into trouble; it's what you know for sure that just isn't so.*

The specious nature of the prosecution's greenhouse gas climate change theory is an inducement to accept it on its face. Once accepted, scrutinization of the theory's "accepted science" becomes more difficult owing to the persistent support it receives in news reports.

It *seems* reasonable, and wrapped in fearful projections of drowning coastal cities, stronger storms, lost winters, worsening floods, longer droughts, and other climate catastrophes, many reasonable jurors might support mitigating action "just in case."

That reaction is all the more reason the jury must follow the evidence and exonerate both fossil fuels and atmospheric carbon dioxide of all charges the prosecution has alleged. Believing a specious theory guarantees an improper verdict based on an *assumed* certainty. An improper verdict is likely to compel costly inappropriate solutions to a problem that simply doesn't exist in the real world. No benefit can be derived from actions motivated by erroneous beliefs that contradict the clear testimony of real world evidence.

Because theory discussion is beyond both the interest and scientific background of most jurors, the defense has relied exclusively on the best available real-world evidence to help the jury examine and understand the actual relationship between changing atmospheric CO_2 and climate change.

Guided by *the scientific method*, the jury examined the evidence that yielded unequivocal compelling testimony that the prosecution's climate change theory is inconsistent with the evidence found in nature.

The prosecution misrepresents a dramatic growth of atmospheric CO_2 since the mid-1940s as being a product of fossil fuel use and couples that misrepresentation with the persistent slight long-term natural warming of global temperatures since the Little Ice Age ended circa 1850. Despite this prosecution effort to mislead the jury, the evidence relentlessly testifies that changing atmospheric CO_2 and global temperature change are strongly *uncorrelated*, a condition that absolutely guarantees there *cannot be a causal relationship* between the two.

The defense has shown the jury there are *at least* two other significant forces that are constantly in play and are known to affect global temperatures:

- Solar activity cycles that impact total solar irradiance,
- Geothermal heat cycles (volcanism and El Niño cycles)

These cycles generally impact a number of years and, should they overlap, can impact climate for years or even decades (e.g., the strong 1998 and 2015–16 El Niño cycles concurrent with a strong solar grand maximum[131]).

The prosecution began this trial nearly four decades ago with assertions that:

- Fossil fuel CO_2 emissions are chiefly responsible for most of the recent growth of atmospheric CO_2.
- Recent atmospheric CO_2 growth is responsible for global warming (climate change).
- Atmospheric carbon dioxide (CO_2) is a strong climate change force.

The defense offered the jury compelling testimony that each of these assertions is unequivocally contradicted by the observed evidence in nature. In accordance with *the scientific method*, such contradictions render each of these prosecution assertions utterly without merit and completely invalid.

Recall from Chapter 1 that the prosecution's theorized relationship between atmospheric CO_2 change and climate change can be summarized as:

> *If global atmospheric CO_2 changes, then global climate will change accordingly.*

In an attempt to support its claims, the prosecution took advantage of two fortuitous late 20th- and early 21st-century warming events to deceive the jury:

- A very strong end of century solar grand maximum.[120]
- Two unusually strong ENSO events (1998 and 2016)

The prosecution exploited the warming caused by these forces to claim something entirely different is responsible for the warming, namely growing atmospheric carbon dioxide. Yet when the jury examined the evidence, it found the prosecution's theorized relationship soundly contradicted by real-world records.

Indeed, the jury discovered the prosecution routinely ignores compelling testimony that unequivocally contradicts its theory:

- From 1941 to 1976, atmospheric CO_2 change and GAST change were just *the opposite* of that claimed by climate change theory. During those 35 years, while atmospheric CO_2 *grew* by +21 ppm, GAST *cooled* by -0.27°C!

- From 1882 to 2018, 137 years of year-to-year atmospheric CO_2 change and GAST change show *just a 46% consistency* with theory and *a 54% contradiction* to theory.
- From 1911 to 2018, 108 years of 30-year atmospheric CO_2 and temperature (climate) changes are *just 50% theory consistent.*
- Sixty-four percent of state all-time high temperature records were set *before* 1950 during *low rates* of atmospheric CO_2 growth.
- Just thirty-six percent of state all-time high temperature records were set *after* 1950 during *rapid rates* of atmospheric CO_2 growth.
- Forty-eight percent of state *all-time low* temperature records were set *after* 1950.
- For 65 years, from 1880 to 1944, global average surface temperature warmed at the same rate of warming as during the 74 years from 1945 to 2018 (0.007°C/year). Yet during the later 74 years, the rate of atmospheric CO_2 growth was *4.55 times greater than* the rate of growth during the earlier 65 years.
- The 114 new NOAA-USCRN land-based temperature-monitoring stations[174] active in the 48 contiguous US states since 2005 show *a net cooling of -0.74°F* between 2005 and 2019 (see Defense Exhibit B) as *atmospheric CO_2 was growing* by about 30 ppm. Once again this cooling demonstrates the lack of any correlation between growing atmospheric CO_2 and surface temperatures in the USCRN records.

The prosecution's climate change theory is so strongly at odds with *the scientific method* requirement that theory match observation that the jury has no alternative but to reject the prosecution's theory as unsupported by *the scientific method* and consequently, completely invalid.

The jury has seen additional compelling evidence that rejects the prosecution's climate change theory, including:

- No correlation exists between atmospheric CO_2 change and global temperature (climate) change over any period of time for which evidence of the two measures exists. Correlation coefficients are in the range of -0.1 to 0.3, *highly uncorrelated* relationships (Chapters 3 and 4).
- The evidence testifies to a chance (flip of a coin) relationship between changes in atmospheric CO_2 and climate, guaranteeing that *no causation is possible* (Chapter 4).
- Current atmospheric CO_2 is extremely low compared with levels typical of the past 550 million years during which severe ice eras occurred numerous times. One ice era featured an extremely sharp temperature *decline* as atmospheric CO_2 *grew* by nearly 500 ppm to *eleven times current levels* followed by a rapid temperature *increase* as atmospheric CO_2 *declined* by 1500 ppm (Chapter 3)! During approximately 83% of the past 550 million years, Earth's concentration of atmospheric CO_2 has been much higher than at present, yet there is no correlation between climate and atmospheric CO_2 concentration (see Chapter 3, Table 3-3).
- Testimony based on year-to-year evidence of atmospheric CO_2 change and global average surface temperature (GAST) change from 1880 to 2018 unequivocally refutes climate change theory. In 54% of those years, atmospheric CO_2 and GAST changed in *opposite* directions (either CO_2 increased and temperature cooled or CO_2 decreased and temperature warmed) (Chapter 4).
- Thirty-year increments of global temperature change (climate) and thirty-year increments of global atmospheric CO_2 change over the 108 years from 1911 to 2018 strongly refute climate change theory (Chapter 4).

- Solar variability since before the Medieval Warm Period has a much stronger correlation with global climate change than does changing atmospheric CO_2 (Chapter 5).
- The family of climate simulation models (CMIP5) based on the prosecution's greenhouse gas climate change theory are known to "run hot" (project much higher temperatures than observed in the real world), offering compelling testimony that the theory behind the models gives far too much power to atmospheric CO_2 as a climate warming force (Chapter 5).
- Since individual state all-time high temperature records began being kept in the 19th century, US fifty-state extreme temperature records show no pattern of warming consistent with the prosecution's climate change theory (Chapter 5).
- Required manifestations of greenhouse gas climate change theory are not observed in real-world measurements, specifically, (1) a tropical mid-tropospheric hot spot of warming does not exist and (2) polar amplification has not been observed (Chapter 5).

The evidence testifying against the climate change theory is overwhelming, unequivocal, and compelling. There is no reasonable basis upon which the prosecution's climate change theory can withstand its persistent contradictions to real-world evidence.

In short, real-world evidence testifies that:

- Global climate change over the past 150 years is entirely caused by natural forces.
- Global climate change has only a chance agreement with global atmospheric CO_2 change.
- Human activity bears no responsibility for global climate change. The use of fossil fuels has no discernible impact on global climate.

According to the testimony of the best available evidence, the prosecution's case is baseless. Climate changes naturally. Severe weather events occur naturally. No observed evidence supports any claim that climate change has any meaningful relationship with changing atmospheric carbon dioxide or methane.

The scientific method invalidates the prosecution's climate change theory upon which its entire case is based.

This trial has given the jury clear and compelling evidence upon which it can confidently affirm that fossil fuel use and atmospheric carbon dioxide are completely innocent of every one of the prosecution's specious climate change charges. There is absolutely no justification for constraining the extraction and use of abundant fossil fuels to supply the energy and transportation requirements of modern civilization.

The jury has also been awakened to the enormous deception underlying the climate change narrative that has frightened school children and caused many others to fear a climate Armageddon on the basis of the prosecution's invalid theory. Jurors have every right to be angered at how many have been duped by the prosecution's scientific fraud the equivalent of a modern-day "Piltdown man" hoax.[121]

Yet as 19th-century statesman Daniel Webster[15] once observed, "Anger never won an argument." Those who may have put their unquestioned faith in the prosecution's dogma should put aside their anger at being duped and become an agent for truth by sharing their newfound understanding with friends and family.

In its quest for climate change truth, the defense exposed the jury to exonerating evidence withheld by the prosecution. The defense reminds the jury that science is never determined by consensus; rather, just as with this trial, the truth that emerges from real-world evidence must be the basis for every sound verdict.

Words from the wise:

> Truth does not become more true if the whole
> world were to accept it; nor does it become less true
> if the whole world were to reject it. (Maimonides)

Albert Einstein on consensus: Upon learning that 100 scientists had co-authored a book (*One Hundred Authors Against Einstein*) challenging his theory of relativity, Einstein reportedly asked:

> *Why 100? If I were wrong, then one*
> *would have been enough.*

EPILOGUE

The anonymous advice above is a caution that advocates who support climate change theory would do well to take to heart.

The evidence examined in Chapters 3-5 is both compelling and sufficient to inform jurors that additional greenhouse gas emissions from any source have no capacity to measurably alter global average temperatures (see Defense Exhibit B, Figure B-2).

The perspective gained from *looking out the window* at the testimony of nature is sufficiently compelling to inform anyone seeking the truth that the prosecution's dubious climate change theory is simply not supported by the observed evidence in nature.

The prosecution's team is certainly devoted to its theory, but theory is by nature, speculation. Theory may or may not reflect reality; it may or may not be scientific. When theory is soundly contradicted by real-world evidence, it becomes unscientific fantasy.

Yet, buttressed by the support of a fawning news media, eager politicians, expanding government agencies, compliant educators, and grant-seeking university academicians, the belief that fossil fuel use is responsible for climate change is championed by the prosecution, claiming growing atmospheric greenhouse gases are a powerful force responsible for recent observed climate warming.

Those who have invested in costly immature alternate energy schemes, those who have taught generations of students a dubious theory, those news media outlets who profit from fearmongering human-caused climate change, together with the many organized extensions of the prosecution's team, will all have strong motives to deny the clear evidence easily seen by *Looking Out the Window*.

For this reason, the revelations described in this book may precipitate negative pushback from those attempting to use this issue for monetary or political advantage. Yet nothing that might be claimed can overcome the powerfully clear testimony of the real-world evidence maintained by US government agencies.

In the end, the compelling unequivocal evidence informs the jury the prosecution's climate change theory is pure speculative fantasy festooned with the trappings of science. Consistently contradicted by real-world evidence, the prosecution's theory is found to be hopelessly invalid.

In the Prologue, the defense asked the jury to

> …objectively view real-world evidence (derived from "looking out the window") from a perspective free of bias and render a fair and just verdict informed by *the scientific method*.

Examined for timeframes ranging from years to more than centuries, contemporary measured records consistently contradict the prosecution's theory. Ice core records spanning tens of thousands to nearly a million years corroborate contemporary evidence. Even geologic evidence spanning more than half a billion years adds further testimony that strongly contradicts the prosecution's theory.

Theory-mandated consequences (tropical mid-tropospheric warming and polar amplification) are missing.

Projections of theory-based climate simulation models testify against the validity of their underlying theory.

The prosecution withholds evidence from the jury that is exculpatory to CO_2. This evidence is found in the testimony of the 114 new USCRN temperature-monitoring stations records showing a net atmospheric cooling of -0.41°C (-0.74°F) over US land surfaces[174] during fifteen years (2005-2019) while atmospheric CO_2 continued steadily growing by 30 ppm!

The evidence shows neither a consistent linkage nor relationship between changing atmospheric CO_2 and changing atmospheric temperature. The testimony of real-world records certifies that climate is not warming in response to growing atmospheric CO_2 as theory advocates claim. Indeed, the relationship revealed is one of pure chance.

The lack of a theory-consistent relationship between temperature (climate) and atmospheric CO_2 renders the question of the extent to which the use of fossil fuels is responsible for atmospheric CO_2 growth entirely irrelevant.

Since climate and climate change are found to be unrelated to either total atmospheric CO_2 or changing atmospheric CO_2, then it really doesn't matter what the source of any CO_2 change might be.

Restrictions on fossil fuels are poorly-conceived non-solutions to a specious theoretical problem whose existence is unequivocally disproven by testimony from real-world evidence showing atmospheric CO_2 plays no detectable role in either contemporary global temperature change or long-term climate change.

Programs like *The Paris Accord* and *The Green New Deal* are costly misguided attempts to control climate change that are doomed to failure for being falsely-based and well beyond both the knowledge and capacity of humans.

What benefit accrues to any nation that would allow its energy resources to be restricted in furtherance of an unjustifiable agenda disguised as an attempt to prevent an impossible human-caused climate change?

While development of cost-effective alternate energy sources is a prudent long-term objective, such actions must be tempered by real-world realities and constrained within the efficiencies demanded by sound economics.

It is pure folly to disrupt access to an abundant, stable, and affordable supply of sufficient energy to power civilization's needs on the basis of invalid speculation contradicted by real-world evidence.

In the prologue, our quest for the truth about climate change began with this advice from Ephesians 4:14:

> We will not be influenced when people try to trick us with lies so clever they sound like the truth. (*Holy Bible, New Living Translation*)

The prosecution alleges its theory of a crime is based on sound scientific principles. Yet the clear testimony of real-world evidence in defense of atmospheric CO_2 unequivocally contradicts and invalidates the prosecution's theory.

Jurors must loosen any bonds to the prosecution's failed climate change dogma if their search for truth is to be successful.

Follow *the scientific method* and the evidence will guide your search to a true and just verdict.

> And ye shall know the truth, and the truth shall make you free. (*John 8:32*)

GLOSSARY OF TERMS

Several terms used extensively in the trial testimony are defined here to provide clarity and avoid confusion or misunderstanding. Terms are listed alphabetically.

Anomaly:

A deviation or inconsistency; in science, the deviation of a value from a long-term average value. Example: An annual temperature anomaly is the difference between the actual temperature and a specific long-term average temperature; the anomaly measures the deviation from that long-term average temperature.

Annual (Year-to-Year) Change, Global Atmospheric CO$_2$:

The global *year-to-year change* (annual change) of atmospheric CO$_2$ (usually in parts per million, ppm). How much does the volume of atmospheric CO$_2$ grow (or shrink) each year? Year-to-year (rather than month-to-month) net CO$_2$ change is examined because annual change smooths the large seasonal variations characteristic of monthly atmospheric CO$_2$.[3, 122] *Note: 1 ppm = 1 in 1,000,000 or 0.000001 by volume, (aka, ppmv); 400 ppm = 0.000400, commonly 0.0004; 400 ppm CO$_2$ means just 0.0004 (or 0.04%) of Earth's atmospheric volume contains CO$_2$.*

Annual (Year-to-Year) Change, Global Average Surface Temperatures (GAST):

The *year-to-year change* (annual change) of Global Average Surface Temperature (usually available in three categories: land, ocean, and merged land plus ocean surfaces). This statistic measures how much GAST changes each year.

Anthropogenic:

From *anthropo-* (of a human being; relating to humankind) and *-genic* (produced by; originating from). Originating from or produced by human activity; human-induced; human-caused.

Anthropogenic CO$_2$ (also referred to as ACO$_2$):

Carbon dioxide emissions produced by human activity.

Figure G-1

Components of Anthropogenic CO$_2$ Emissions
(2014)
- Land Use Emissions (13%)
- Fossil Fuel Emissions (82%)
- Non-hydraulic Cement Production Emissions (5%)

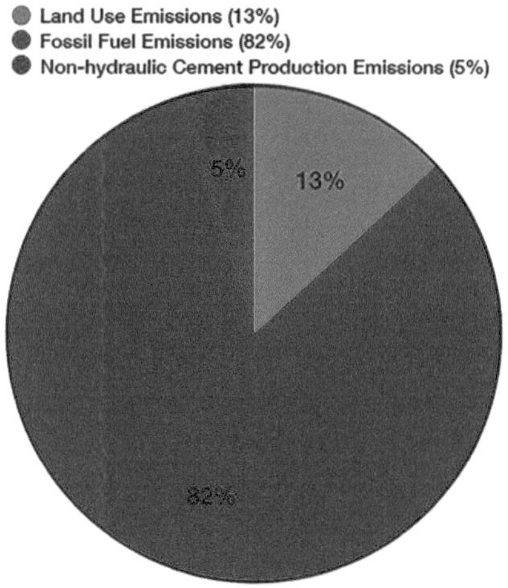

Examples of ACO$_2$ sources: fossil fuel combustion, land use, and nonhydraulic cement production. See Figure G-1 at right. Fossil fuel combustion (the dominant source at 82% of all anthropogenic emissions in 2014), land use (next most significant at 13% of anthropogenic emissions), and nonhydraulic cement production (at 5% of total anthropogenic emissions; note that cement production emissions are reabsorbed during cement curing yielding *a net of no change* in atmospheric CO$_2$ after curing). Were it not for fossil fuel emissions, there would be no case against ACO$_2$ or carbon dioxide.

Correlation Coefficient ("r" or "ρ"):

Whenever *a high degree of consistency* exists between two sets of data (i.e., changes in one are paired with responsive changes in the other), they are considered correlated. When no appreciable consistency exists, the data are considered uncorrelated. Such relationships are measured by a statistic known as a correlation coefficient (usually denoted by "r" or "ρ"). [Note: "ρ" is the Greek letter *rho.*]

If two data sets are correlated, *there might exist a causal relationship* between the two, but *no causal relationship is guaranteed* (the relationship could be *by chance*). An example of a chance correlation is comparing the price of bread in Britain with sea levels in Venice, Italy. Over the past two centuries, both the price of British bread and the sea level in Venice have risen leading to a high correlation between the two. Yet clearly, the price of British bread is not dependent on sea level in Venice, and neither is the sea level in Venice dependent on the price of British bread! While correlated, *there is no causation.*

When teaching statistics, the difference between correlation and causation is often illustrated by the *Stork and Baby Trap:*[123]

It was observed in an Alsatian village that the number of new babies in the village is highly correlated with the number of nesting storks. This might lead one to conclude that storks do, indeed, bring babies. But, of course, we know that isn't the case. So why is there such a high correlation? The simple truth is in the fact that the number of new babies in the village rose as the village population increased. As the population increased, more homes were built, increasing the number of chimneys for storks to nest on, which drew more storks to nest in the village. In this example, while there is no direct causation between storks and babies, both the number of storks and the number of babies are responsive to the village population.

Correlation cannot prove causation, but causation will create correlation; lack of correlation establishes lack of causation. If A causes B, then A and B must be correlated. If A and B are uncorrelated, then A cannot cause B. Correlation between two data sets is measured by a numerical statistical relationship known as the correlation coefficient, which value ranges from -1 to +1.

A correlation can be either positive (the direction of change for both data sets is the same; if one goes up, the other goes up) or negative (the direction of change for both data sets is opposite, if one goes up, the other goes down). A correlation coefficient of zero means there is no relationship whatsoever between the two datasets (for example, two sets of random numbers). Datasets with correlation coefficients between 0.3 and -0.3 are considered *highly uncorrelated*. Datasets with correlation coefficients *between 0.8 and 1.0 or -0.8 and -1.0 are positively and negatively correlated, respectively.*

If two data sets are uncorrelated, there cannot be a causal relationship between the two.

The nature of the greenhouse gas climate change theory requires that, in the absence of a strong mitigating force, when atmospheric CO_2 changes, temperature must consistently change accordingly. There is no provision for CO_2 changes to not be reflected in corresponding temperature changes without acknowledging there are other, more powerful, forces working independently of and dwarfing any effect from atmospheric CO_2. A disconnect between annual changes in global atmospheric CO_2 and annual changes in global average temperature is a strong indication that something other than atmospheric CO_2 dominates changing climate.

El Niño Southern Oscillation (ENSO):

An irregularly periodic variation in winds and sea surface temperatures over the tropical eastern Pacific Ocean, affecting climate of much of the tropics and subtropics. The warming phase of the sea temperature is known as El Niño and the cooling phase as La Niña.[124] ENSO oscillations (between El Niño, neutral, and La Niña) are known to affect global circulation patterns and climate. Evidence of ENSO cycles go back to at least 1876, lending credibility to a relatively new theory that likens the trigger for ENSO cycles to the cycles of geysers.[24, 52] It is theorized porous layers along the sides of the Mariana Trench periodically become saturated with superheated water that is expelled in a grand eruption of very warm water that rises to the surface of the western Pacific Ocean and migrates over the Pacific Ocean surface to the east coast of South America.

GAST (Global Average Surface Temperature):

Global Average Surface Temperature (usually available in three categories: land, ocean, and merged land plus ocean surfaces).

Intergovernmental Panel on Climate Change (IPCC):

An organization established under United Nations auspices to assess the consequences of global climate change. The IPCC was tasked to stabilize "greenhouse gas concentrations in the atmosphere at a level that would prevent dangerous anthropogenic [i.e., human-induced] interference with the climate system."[125] This tasking *assumes a causation that has not been established*, that is, that global climate change is caused by humans using fossil fuels for energy. A proper tasking would have been to unequivocally establish (if possible) the cause of observed climate change over the past 170 years using *the scientific method*. Once established (not assumed), then and only then would it be appropriate to address whether any actions by humans are even capable of altering future climate. Instead, a climate change cause was assumed, and the panel was directed to operate accordingly. By any reasonable standard, that is not a scientific directive, it is a political one.

Isotope:

An isotope is an alternate form of an element differing only in the number of neutrons but not in chemical properties.[126] Examples: carbon has three isotopes, ^{12}C (standard, 99%), ^{13}C (1%), ^{14}C (trace).

NAO:

North Atlantic Oscillation, an atmospheric phenomenon tied to the sea level pressure differential between the Icelandic Low and the Azores High. These cycles affect the course of westerlies and storm tracks across the North Atlantic. (Wikipedia)

pH:

A pH unit is a measure of acidity or alkalinity ranging from 0-14. A pH of 7 is neutral, neither acidic nor alkaline. A pH less than 7 is acidic. A pH greater than 7 is alkaline. Acidification is any relative shift in pH to a lower value; alkalization is any relative shift in pH to a higher value.[127] Acidification (lowering pH) of an alkaline solution can remain alkaline; alkalization (raising pH) of an acidic solution can remain acidic.

Solar Insolation:

Solar insolation is the cumulative effect of instantaneous solar irradiance; the amount of solar energy that is available at a given location, per unit of area and time. Solar insolation is typically measured in kilowatt-hours per square meter per day (kWh/m²/day).[128] Absent any mitigating factors, a prolonged increase (or decrease) in global solar insolation will lead to climate warming (or cooling).

Solar Irradiance:

Solar irradiance is the instantaneous solar power per unit area received in the form of electromagnetic radiation. Irradiance on the Earth's surface depends on the tilt of the measuring surface, the height of the sun above the horizon, and atmospheric conditions.[128] Solar irradiance is variable throughout the day. Absent any mitigating factors, a prolonged increase (or decrease) in global solar irradiance will lead to climate warming (or cooling).

Solar Grand Maximum and Solar Grand Minimum:

The sun experiences 11-year cycles of solar activity that includes sunspots, solar flares, and "large bursts of plasma known as coronal mass ejections (CMEs), or solar storms [which are m]ore common during the active period of the cycle known as the solar maximum. CMEs have a stronger effect than the standard solar wind."[129]

"Sunspots are strongly magnetized, and they crackle with solar flares—magnetic explosions that illuminate Earth with flashes of X-rays and extreme ultraviolet radiation. The sun is a seething mass of activity. Until it's not. Every 11 years or so, sunspots fade away, bringing a period of relative calm. 'This is called solar minimum,' says Dean Pesnell of NASA's Goddard Space Flight Center in Greenbelt, MD. 'And it's a regular part of the sunspot cycle.'"[130] Solar maxima occur during the peak of solar activity in the 11-year cycle; solar minima occur during the valley of solar activity in the 11-year cycle.

Solar maxima and minima have their own cycles based on a much more complex cycle. A peak solar maxima is known as a solar grand maximum; a valley of solar minima is known as a solar grand minimum. A solar grand maximum occurred recently. "The modern Grand maximum (which occurred during solar cycles 19–23, i.e., 1950–2009)" is "a rare or even unique event, in both magnitude and duration, in the past three millennia" (3,000 years).[131, 132] Many solar scientists believe a solar grand minimum will occur in the very near future and solar maxima may be significantly less intensive for decades to come.

Specious:

Superficially plausible, but actually wrong.

Symbols and Abbreviations:

Symbol	Meaning	Abbrev.	Meaning
>	greater than	ACO_2	anthropogenic carbon dioxide
<	less than	CO_2	carbon dioxide
>>	much greater than	GHG	greenhouse gas
<<	much less than	Gt	gigatons (ex. 321 Gt = 321 billion metric tonnes)
~	approximately	my	million years
=	equal to	mya	million years ago
Δ	change in (Greek uppercase *delta*)	ppb	parts per billion
ρ	correlation coefficient (Greek lowercase *rho*)	ppm	parts per million (often interchanged with with ppmv)
∴	therefore	ppmv	parts per million of volume
		tr	trace or trivial, virtually nil, minuscule
		TSI	Total Solar Irradiance

The Scientific Method:

A method for *testing a theory or hypothesis* by one or more of the following:

a. evaluating its conformity with laws of science (is any scientific law violated?)

b. testing by experimentation (is it compatible with experimental results?)

c. real-world evidence to assess conformity with reality (does it conform with the observational evidence found in nature?).

If a theory or hypothesis *fails any one of these tests*, it *must be* rejected or altered.

To have the greatest confidence in a theory, *it should survive scrutiny of all three tests* of *the scientific method*; however, failure of any one of them, even if the theory survives the other tests, requires the theory to be discarded or revised.

When addressing the question of a theory's validity, neither the complexity of the theory's scientific derivation nor the number of scientists (consensus) who endorse the theory are relevant factors. If a theory is contradicted by *any one* of the tests of *the scientific method*, the theory is, by definition, *invalidated*. Consequently, *the scientific method* is a valuable tool to *invalidate erroneous theory*. On the other hand, *the scientific method* can lend support to a theory if (1) the theory does not violate any known scientific principles, (2) the theory conforms with experimentation, and (3) the theory is consistent with real-world evidence (ex. Einstein's theory of relativity). Note that, due to its very nature, testing a theory using *the scientific method cannot completely validate* a theory (i.e., prove the theory); at best it can only *fail to invalidate it*.

Jurors should understand *the scientific method* as it is the most legitimate method for a truly competent scientific examination of speculative hypotheses such as climate change theory. Consequently, *the scientific method* is referenced throughout this trial.

Noted Scientists on *the scientific method*:

Albert Einstein: "No amount of experimentation can ever prove me right; a single experiment can prove me wrong."

Nobel Laureate Richard Feynman: "It doesn't matter how beautiful your theory is, it doesn't matter how smart you are. If it doesn't agree with experiment, it's wrong."

Paraphrasing Feynman: "It doesn't matter how beautiful your theory is, it doesn't matter how smart you are. If your theory doesn't agree with observed evidence in nature, it's wrong."

Troposphere:

The part of Earth's atmosphere that extends from its surface to about seven miles altitude (~37,000 feet) at the border with the stratosphere. The troposphere is that part of the atmosphere in which most weather occurs. Temperature typically decreases rapidly with altitude.

DEFENSE EXHIBITS

Additional Evidence for the Jury

DEFENSE EXHIBIT A

Disinformation in the News—Reliable Alternate Information Sources

Disinformation: Prosecution Dogma in News Reports

The compliant news media are devoted to the human-caused climate change narrative. Regardless whether a legitimate basis exists, fear-mongering sells. This mindless devotion to the climate change narrative is evident in virtually every news article that references climate change.

Some of the more egregious examples of these reports:

- *USA Today*, March 15, 2017, *Climate Change Is Making Us Sick, Top U.S. Doctors Say*:[133]

    The burning of fossil fuels—gas, oil and coal—to power our world releases greenhouse gases such as carbon dioxide and methane into the Earth's atmosphere, warming the planet to levels that cannot be explained by natural climate cycles.

Defense statements: USA Today's assertion that post Little Ice Age warming "cannot be explained by natural climate cycles" is baseless nonsense. Blatantly false statements put into "news" stories without a shred of supporting evidence serve only to expose *USA Today*'s ignorant bias.

Warm and cold periods have been characteristic features of uncounted interglacials during an aggregate 350 million years of the past 3.5 billion years during which Earth has experienced seven separate ice eras.

The testimony of real-world evidence clearly reveals that the prosecution's theory and repeated claims that fossil fuels are responsible for climate warming are categorically invalid.

- *USA Today,* June 29, 2017, *Who Will Pay Most for Climate Change? South Will Be Biggest Loser:*[134]

  The research…found that *if fossil fuels continue to pour carbon dioxide* into the atmosphere the U.S. will divide further into a country of haves and have nots. "If we continue on the current path, our analysis indicates it may result in the largest transfer of wealth from the poor to the rich in the country's history."

Defense statements: No credible evidence exists that atmospheric CO_2 growth during the past 3.5 billion years has ever led to climate warming. Theories aren't evidence. Chemical analyses certify the insignificant atmospheric accumulation of fossil fuel emissions[31] *cannot be causing* the observed growth of atmospheric CO_2 (shown to be irrelevant anyway).

- *USA Today,* March 26, 2017, *How Trump's Energy Order in March Affects Jobs, Fuel Prices:*[135]

  Former President Obama's Clean Power Plan in 2015 *curbed emissions at power plants,* a*iming to combat climate change.* Trump wants to roll back the plan, which had slashed allowable carbon pollution.

Defense statements: No credible evidence exists to support the view that curbing CO_2 emissions will have any detectable impact on *climate change.* No evidence exists to suggest humans can do anything to impact future climate change. The most intelligent course would be to prepare for inevitable climate change that will occur regardless of human activity. Carbon dioxide is not pollution—it is the gas of life.

- *CNN,* September 15, 2017, *Yes, Climate Change Made Harvey and Irma Worse:*[136]

  …the consensus among scientists is that the effects of climate change, such as rising sea levels and warmer oceans, made those storms far more destructive than they would have been in previous decades.

Defense statements: First, the "consensus" claimed by *CNN* does not exist; even if it did, science is never determined by consensus; the claim of consensus is based on shoddy work designed to create the illusion of a wished-for consensus. A "consensus among scientists" *CNN* contacted does not constitute an authoritative source, particularly when the consensus runs contrary to recorded evidence and well-established science.

Second, a *CNN* graphic shows sea level rising 3.5" over the past 25 years. The implication is that storm surges are "far more destructive." Really? Consider a *modest* 10' storm surge associated with a hurricane (for perspective, Katrina's 2005 storm surge was 27.8', Camille's 1969 surge was 22.6', Charley's 2004 surge was 18', Ike's 2008 surge was 17').[137] Is a 10'-3.5" (123.5") storm surge *really* far more destructive than a 120" storm surge? A 3.5" increase in Katrina's storm surge would have amounted to a minuscule 1% increase! Are the people at *CNN* nuts or just plain stupid?

Third, has *CNN* made any *serious* attempt to *understand why* sea levels might be rising? Not likely. Does *CNN* understand that sea levels have been slowly rising since the 19th-century end of the Little ice Age? Probably not.

Fourth, *CNN*'s graphic is misleading as it does not reflect *global* sea level rise. This is yet another of many unsupportable *CNN* claims (fake news seems an apt description). And no, there isn't a scintilla of evidence that human activity had anything to do with *any* recent hurricane (Harvey, Irma, Katrina, or any other).

Finally, *CNN* routinely engages in such fearmongering, serving up *mindless drivel* that cannot withstand even the most elementary scrutiny. *When it comes to climate change news, CNN* reports are eminently deserving of the label "fake news."

- *USA Today*, April 23, 2018, *California's Wild Extremes of Drought and Floods to Worsen as Climate Warms*,[138] claiming:

> Over the past couple of years, California lurched from its worst drought ever to disastrous, record flooding.
>
> Now, a new study suggests the frequency of these rapid, year-to-year swings from extreme dry to wet conditions—which the study authors dub 'precipitation whiplash events'—may become more common in California's future as a consequence of man-made global warming.
>
> Of particular concern would be a repeat of the infamous 1862 flood that likely killed thousands of people…
>
> Here's the science behind how it works: The amount of water vapor in the atmosphere increases rapidly as the atmosphere warms. This big increase in water vapor can lift the 'ceiling' on extreme rainstorms and snowstorms, leading to a greatly increased potential for very wet events.
>
> Global warming could also make the dry times even drier. In addition to the water vapor increase, climate change is likely to change prevailing wind and storm track patterns over the Pacific in a way that favors both wetter California winters in some years and drier winters in other years…

Defense statements: This story is teeming with speculation and inappropriate phrases.

First, the claims "worst drought ever" and "record flooding" are meaningless without reference to a time frame. "Worst drought and flooding since official US Weather Bureau records began in 1870" would have been appropriate (if true). But to write "worst" and "record" without reference to a timeframe is meaningless hyperbole (and *USA Today* must know that). As written, this is nothing but fearmongering.

Second, to write that extremes of dry and wet conditions "may become" more common is pure speculative nonsense added without any sound foundation being provided. They also *may not* become more common.

Third, attribution of drought and flood frequency to "man-made global warming" is nonsense and entirely inappropriate since there isn't a scintilla of real-world evidence that man has any capacity to discernibly impact either weather or climate on a global scale. A news agency evidently more concerned with scaring readers for profit than being truthful, *USA Today* simply tosses that phrase in to try to legitimize its speculation.

Fourth, the "infamous 1862 flood that likely killed thousands" is referenced, evidently because it likely killed people, yet haven't we been told that this kind of flooding would be *unprecedented* were it not for "man-made global warming"? So which is it? Furthermore, the cited precedent flood occurred *eight years before* official US Weather Bureau records were begun in 1870, so the true extent of the rainfall that caused the flood is not part of the official record. If *USA Today* is so certain that global warming had a hand in the recent flood, what do they supposed caused the 1862 flood? Also, the phrase "*likely killed*" is meaningless. For example, if there isn't a good record of thousands being killed, the author could just as easily have written "hundreds" in place of "thousands" given the qualifier "likely." If natural forces could create such flooding in 1862 without human help, then those same natural forces are perfectly capable of naturally producing such flooding today and in the future.

Fifth, the discussion about how "the science" works is grossly misleading. The science we really know has an explanation for what causes California's cycles of drought and rainfall, just as it knows what causes midwestern US tornado cycles and Gulf and Atlantic coast hurricane cycles. There is no evidence that the offered science has ever actually had a hand in any rainfall increase anywhere in California as the explanation is simply theoretical. The claim "water vapor…increases rapidly as the atmosphere warms" is an oversimplification that suggests there is no such thing as a hot dry desert climate! Tell that to the people living in the desert southwest of the USA or the Sahara. This is classic "having your cake and eating it too" rationale. Warming creates more water vapor and warming makes it drier! Pretty neat trick, right? In the nontheoretical real world, for any two inland locations on the same latitude, the location with higher water vapor content (greater humidity) is much more likely to be the *cooler* location; the drier is more likely to be the *warmer*. The

false message conveyed by *USA Today*: Warming is evil and it's all humanity's fault! Again, this claimed science is both naive and misleading, not a particularly uncommon trait for *USA Today*.

Sixth, the article claims "warming could" make dry times even drier, yes, and it "could" have no such impact. Yet again, another misleading generalization unsupported by Earth's long history of climate and atmospheric CO_2 and contradicted by every hot dry climate on the planet!

Finally, prevailing wind patterns are often influenced by ENSO cycles that have nothing to do with atmospheric CO_2. This is yet another gratuitous example of fake news where every possible opportunity is taken to suggest "man-made global warming" is both real and responsible for *perfectly natural* variations in climate and weather events.

- *USA Today*, October 13, 2018, *Global Warming Suspected of 'Supercharging' Michael to a Monster,* claiming:

  Although random weather patterns certainly played a role, the warm waters in the Gulf have a *"human fingerprint"* of climate change, according to National Oceanic and Atmospheric Administration climate and hurricane expert …

Defense statements: This story appeared three days after Hurricane Michael made landfall on October 10, 2018, along the Florida panhandle. The NOAA "climate and hurricane expert" claims that global warming (code for human-caused global warming) was responsible for the size and strength of Hurricane Michael.

First, as jurors are now well aware, and as the climate and atmospheric CO_2 records clearly reveal, humans have no impact whatsoever on global climate (a.k.a., global warming).

Second, the NOAA expert claim was based on the notion that Gulf of Mexico waters were warmer than normal due to global warming caused by humans (the human fingerprint claim) and that such warming was responsible for supercharging Hurricane Michael into a monster storm. Yet when this NOAA expert was contacted and politely asked to explain the basis for his claim that the small pocket of warm water in the northeast Gulf was a consequence of *any human activity*, he had nothing to say. Historic records show that such warm pockets of water in the Gulf of Mexico in early October are fairly routine.

Third, since the climate change evidence clearly demonstrates that climate warming is neither unusual nor anything other than a consequence of natural processes, this expert claim cannot be validated.

Fourth, while Michael was a minimal category 5[†††] storm, it was of relative short duration, rapidly dwindling from a category 5 level to a modest tropical storm *in the space of just 14 hours* (midday October 10:00 to 2:00 a.m. October 11). While the devastation in the eye of the storm as it made landfall had all the appearance of the devastation from a large tornado, *CNN*'s use of the terms *monster* and *supercharged* is arrogant.

Finally, this type of article quoting a NOAA expert making off-the-cuff statements that he cannot possibly justify is typical of the unquestioning role played by a compliant media that appears all too willing to give credibility to the *grand fraud of human-caused climate change (a.k.a., "global warming")*.

As the late Prof. István Markó observed:[139]

> The fact that masculine fertility decreases; the fact that birds' wings shrink; the fact that a shark showed up in the North Sea; absolutely anything is likely to be connected to climate change if one displays enough intellectual dishonesty.

- *USA Today*, May 15 (update), 2018, *Supercharged by global warming, record hot seawater fueled Hurricane Harvey,*[140] claiming:

> Record warm water in the Gulf of Mexico fueled the historic rainfall from Hurricane Harvey last August, according to a new study, which also found man-made climate change was partly to blame.

> The heat in the Gulf of Mexico last August, just prior to Hurricane Harvey, was the highest ever recorded, scientists said.

> "Harvey could not have produced so much rain without human-induced climate change," the study said. More troubling, this could be a harbinger of future storms." As climate change continues to heat the oceans, we can expect more supercharged storms like Harvey," according to Trenberth.

[†††] Based on post-storm analysis of data, on April 19, 2019, Michael was upgraded by NOAA from a strong category 4 storm at landfall to a weak category 5 storm.

[The lead author of the study is Kevin E. Trenberth, a long-time advocate for the prosecution's climate change theory.]

Defense statements: This article conflates a smattering of truth with massive speculation in order to fear-monger. Trenberth has no evidence whatsoever that Gulf of Mexico seawater temperatures (historically much warmer than oceans) are any warmer as a consequence of fossil fuel emissions. He just says it. The evidence belies Trenberth's speculation.

Hurricane Harvey's greatest destruction was caused by the prodigious rainfall it produced in Texas. A record amount that exceeded sixty inches (five feet!) of rain fell. It was this rainfall that caused the destructive flooding well beyond anything humans had ever experienced in that area. However, the cause of that flooding had *nothing to do with climate change or global warming!* The flood-producing rainfall was created by mid- and upper-level atmospheric steering conditions that locked Hurricane Harvey in place for days, allowing heavy rainfall to produce the record flooding.

Furthermore, given the evidence of the trivial contribution[31–34] of fossil fuel CO_2 emissions to total atmospheric CO_2 (2% to 3%) and the compelling evidence that total atmospheric CO_2 *has no impact* on global climate change, the mere notion of "human-induced" or "man-made" climate change is bizarre and unsupportable. *Climate deception* is the thread commonly used to weave these reports.

The news stories cited above are based on a level of intellectual dishonesty that inevitably is reduced to pure speculation by uninformed and/or deliberately misleading writers and/or dishonest interview subjects. The only certainty is real-world evidence that carbon dioxide plays no discernible role in climate change. Fakery or quackery might be more appropriate terms than science or news when it comes to such stories.

When misleading fake claims such as the above are routinely served up to an unsuspecting public, unfortunately some people may find them believable. In fact, they are just another example of fake news that pervades what is perhaps inappropriately referred to as news media but which appears to be morphing into propaganda media or fake news media that operates with a heavily biased agenda.

On the other hand, record-breaking cold in November 2019 brought the hopes of an early ski season to northwestern New Jersey. According to Hugh Reynolds, Vice-President of Marketing for the Mountain Creek ski resort, "If we are able to open on Saturday, it would be historic… We've never offered any kind of skiing here prior to Thanksgiving, so [we are] taking advantage of this record-cold temperature…"

- *NJ Herald,* November 14, 2019, *Record lows bring early snow:*

  Nick Stefano, a Wantage resident who has weather records for Sussex County dating back to 1893, said a low temperature of 11 degrees was recorded before dawn Wednesday near the High Point Monument in Montague while a station in Walpack read 13 degrees. Both locations broke the previous state record for the date of 14 degrees— set in Layton in 1950—while a Wednesday reading in Sandyston tied the mark.

  [According to Stefano]…the all-time record low statewide for the month of November…was also set in Layton, at minus 5 degrees on November 26, 1938.

According to Reynolds, Mountain Creek was open from December 14 to April 7 last winter. The 115 consecutive days marked the longest the resort has ever been open in one season, and employees are optimistic that Saturday's projected start date is a positive sign of things to come.

"Last year was a historic season for us, and we're starting off with a potentially historic opening [this year]", Reynolds said. "We're super excited about what that means for the season ahead."

Defense statements: This is another notable example of the many contradictions to the climate change narrative that claims Earth faces an "existential threat" from dramatically warming climate.

Jurors might wonder why the following report, published just one day after the "wild extremes" story (the fourth), didn't seem to warrant the attention of the "the usual suspects" in US news media:

* *Express* (UK), April 24, 2018, *Climate Change Is 'Not as Bad as We Thought' Say Scientists,*[141] claiming:

Climate change is likely to be markedly less severe than forecast, a study claimed yesterday.

It predicted that the impact will be up to 45 per cent less intense than is widely accepted.

It forecast that future warming will be between 30 per cent and 45 per cent lower than suggested by simulations carried out by the UN's Intergovernmental Panel on Climate Change (IPCC).

Nicholas Lewis, one of the authors, said "future warming is likely to be substantially lower than the central computer model-simulated level projected."

…scientists at Plymouth University said average winter wave heights along the Atlantic coast of Western Europe have been rising for seven decades.

Defense statements: If this has been ongoing for seven decades (since the 1940s) it cannot be attributed to any clearly identifiable human activity and is likely to be little more than the impact of perfectly normal modern warming after the Little Ice Age ended in mid 19th-century.

Given the demonstrated interest in this subject by *USA Today* and *CNN*, jurors might have expected this story to be widely reported, but not so much as a peep about this story from either source, despite it having been widely reported in the UK's *Telegraph* and *Daily Mail* and by many online news websites.

Giving Credit Where Credit Is Due

- *USA Today*, January 29, 2019, *Chicago Will be Colder than Antarctica This Week; Record-Breaking Winter Freeze to Hit 250 Million across US*:

Weather.us meteorologist Ryan Maue predicts that a whopping 250 million people in the continental U.S. will experience a freezing temperature by week's end. And 90 million will see below-zero [F] degrees.

In Chicago, schools were already ordered shut for Wednesday due to wind chill temperatures that AccuWeather meteorologist Elliot Abrams said could make it feel like minus 50 to 60. That would be colder than Antarctica.

Minneapolis is forecast to reach minus -54 Wednesday morning. In Wisconsin—a state familiar with brutally cold winters—parts of the state were hit with a foot of snow or more Monday. School districts, government agencies and businesses were poised to remain closed deep into the week due to the cold, and Gov. Tony Evers declared a state of emergency because of the snow and cold.

"The intensity of this cold air, I would say, is once in a generation," said John Gagan, a National Weather Service meteorologist based in Sullivan, Wisconsin.

Defense statements: USA Today ran this story because it is simply too big to be avoided. It should have been the headline story. Nevertheless, *USA Today* did report the record-setting Arctic blast that, according to Al Gore and other climate alarmists, was simply not possible in 2019. Incidentally, in January 2019 Illinois set its all-time record cold temperature. Global warming?

When stories fit the climate change narrative, man-made climate change is blamed; when stories contradict the narrative, they either go unreported or there is no mention of man-made climate change, nor is there any doubting the climate change narrative! That's deliberate disinformation.

Apparently, most G20 Summit attendees accept without scrutiny the self-serving "studies" regularly churned out by prosecution scientists. At the 2017 and 2018 G20 Summit meetings, without any concern for the consequences and without questioning the basis for doing so, member nations simply fell into line and reaffirmed their politically-correct commitment to abide by the fossil fuel emissions restrictions of the *Paris Agreement*.

Yet that agreement assumes at its core the erroneous claim that fossil fuel emissions are the principal cause of atmospheric CO_2 growth that, in turn, is the dominant cause of climate change. In fact, the record clearly shows atmospheric CO_2 has no detectable impact on global climate, so it is very clear that fossil fuel use can have no impact either. Therefore, any costly agreement to try to lower atmospheric CO_2 by restricting CO_2 emissions is senseless folly and cannot have a discernible impact on climate.

The real irony is that even if the *Paris Agreement* restrictions were achieved, no detectable impact on climate change would be derived from the agreement's assault on fossil fuel use. The only certainty will be the negative impact on the economies of participating nations and their aggregate billions of citizens.

We now know for certain (Chapters 3–5) that there is no meaningful connection between the use of fossil fuels and either global atmospheric CO_2 growth or global climate change.

The lone G20 exception to compliance with the *Paris Agreement* is the United States. Under President Donald J. Trump, the US wisely withdrew from the *Paris Agreement*. That withdrawal prompted expressions of shock and outrage among the leading prosecution-duped proponents of the view that carbon dioxide (the gas essential for life) is a pollutant.

At every opportunity, the usual suspects cited above improperly demonize fossil fuels as being responsible for creating potentially catastrophic human-caused climate change (a.k.a., global warming, anthropogenic global warming, AGW, or catastrophic anthropogenic global warming, CAGW).

Jurors might reflect, if the climate change issue is so callously misrepresented, what other issues are these purveyors of "news" misrepresenting? President Trump did exactly what any well-informed conscientious president should have done in the best interests of his nation. For his courageous action he is vilified by dupes who have been misled by the examples cited above.

The good news is that there is no need to monitor carbon footprints or chatter on about carbon pollution and carbon offsets or create carbon credit regulations. The first is entirely irrelevant; the second ignores the reality that carbon dioxide is, has always been, and will always be a life-essential gas without which no life on Earth could exist; and the last two would have no effect other than to disrupt normal commerce and economic vitality while making undeserving carbon credit brokers billions in profits.

Information: Reliable Online Resources

Highly recommended for those with a science background are Dr. Tom V. Segalstad's articles[142] and website[143] that explore many of the difficulties with the prosecution's expedient unscientific assumptions relating to carbon dioxide. For those who do not have a science background, Segalstad's website contains very useful information about climate change.

Also highly recommended and one of the leading sites for climate change news is Anthony Watts's *Watts Up With That?* (https://wattsupwiththat.com).[144]

There are many independent websites—many of which are owned and operated by scientists with pertinent expertise, where useful information and articles can be found that objectively address the climate change issue.

Listed below is a partial selection of some of the many websites where jurors can find timely, useful material in articles and papers that discuss a broad range of climate change issues (not listed in any particular order):

- Dr. Tom V. Segalstad's *CO_2 Web*[143]
- Anthony Watts' *Watts Up With That?*[144]
- *Committee for A Constructive Tomorrow (CFACT)*[145]
- *Principia Scientific International*[146]
- Hans Schreuder's *I Love My Carbon Dioxide*[147]
- Francis Menton's *Manhattan Contrarian*[148]
- *Climate Change Dispatch*[149]

- Mark Morano's *Climate Depot*[150]
- Joanne Nova's *JoNova* (http://www.joannenova.com.au/)
- Jennifer Marohasy's Jennifer Marohasy (https://jennifermarohasy.com/)
- Joe D'Aleo's *ICECAP*[151]
- Tony Heller's *Real Climate Science*[152]
- Judith Curry's *Judith Curry*[153]
- *Science Matters*[154]
- Ole Humlum's *Climate 4 You*[86]
- Dr. Roy Spencer's website (http://www.drroyspencer.com/)
- James Delingpole's *Delingpole World* (https://delingpoleworld.com/)
- Pierre L. Gosselin's, *No Tricks Zone* (http://notrickszone.com/)

Some of these sources are more technical than others and may not appeal to those lacking a scientific background.

There are many more outstanding websites in addition to those listed above. Many will carry articles that all jurors can appreciate, and these might help jurors acquire and maintain a better perspective on the climate change issue.

One thing is abundantly clear—no matter how powerful the evidence that contradicts the human-induced climate change dogma, the prosecution will never stop its war on carbon dioxide.

It's up to jurors to be sufficiently well-informed to stand in the way of ludicrous expensive folly, e.g., a carbon tax; mandatory regulation of carbon "footprints," carbon credits, and credit exchanges; *the Green New Deal*; and nonsensical agreements such as *The Paris Climate Agreement*[155] (a.k.a. *the Paris Accord, Paris Agreement,* and *Paris Climate Agreement*).

It is also vital for jurors to ensure public education institutions *be required to justify their curricula*, particularly if unproven, unscientific dogma (e.g., human-caused climate change that is relentlessly refuted by real-world evidence) is being offered to students as settled science when, in fact, it is unscientific nonsense lacking any meaningful support in nature. Remember, only *scientific laws* are settled—*everything else* is either theory or pure speculation.

Question politicians when they seek election to state or federal office. Do they mindlessly follow the prosecution's dogma, or have they seriously considered the issue by reviewing the mountain of real-world evidence that categorically invalidates the prosecution's climate change theory.

It is vital that poorly-informed politicians who support fantasy over reality not continue to be a part of the problem. Those who blindly ignore the truth and support carbon restriction regulations must be removed from office at the earliest opportunity.

Reality is not a matter of opinion. It is a matter of science and the testimony of real-world evidence to what happens and has happened in nature for the past half billion years. *The scientific method* strongly rejects the prosecution's climate change theory as having been invalidated by the testimony of real-world evidence. The science rejects IPCC dogma just as certainly as IPCC's climate simulation models refute the IPCC climate change theory upon which they are based (Chapter 5, When Computer Models Get It Wrong, Question Their Basis).

There are many more resources addressing this issue both in print media and online.

Be skeptical. Remember, *skepticism* is the essence of *the scientific method*.

Question sweeping claims made without proof that rely on appeals to authority, assumption-based speculation, misrepresentation, and dogma. This is particularly true when the authority cited is more perception than reality.

When offered a theory or an opinion on any subject, do your own research into its veracity. Question whether the assertions made can withstand scrutiny of the evidence.

Additional Topics Supporting Defense Testimony

Do Anthropogenic CO_2 Emissions Cause Ocean Acidification?

A frequently heard claim: *ocean acidification is caused by anthropogenic (ACO₂) emissions.*

How does a typical juror interpret that statement? Would a juror be led to believe that, thanks to fossil fuels, oceans are acidic and becoming even more acidic?

If so, that juror is being misled.

What is really meant by ocean acidification?

The degree to which any solution is either *acidic* or *basic (alkaline)* is indicated by its pH.

Every solution has a pH. A pH of 7 is neutral (i.e., neither acidic nor alkaline, basic).

What is pH?

A pH unit is a measure of acidity ranging from 0-14. The lower the value, the more acidic the environment. Becoming more acidic is a relative shift in pH to a lower value.[127]

According to NOAA:

When carbon dioxide (CO_2) is absorbed by seawater, chemical reactions occur that reduce seawater pH, carbonate ion concentration, and saturation states of biologically important calcium carbonate minerals. These chemical reactions are termed "ocean acidification"[156]

181

Based on the above, the prosecution alleges:

> About 30% of the emitted anthropogenic CO_2 is absorbed by oceans, which reaction reduces carbonate ion concentration, lowering seawater pH.

But is this the whole story? Is this the proper perspective? Is the term *ocean acidification* a correct yet deceptive term for such shifts in pH?

Consider the following testimony for the defense (*emphasis* added):

> Acidic and basic (or alkaline) are two extremes that describe a chemical property. A substance that is neither acidic nor alkaline is neutral.

> The pH scale measures the acidity or alkalinity of a substance. The pH scale ranges from 0 to 14. A pH of 7 is neutral. A pH less than 7 is acidic. A pH greater than 7 is alkaline.

> The pH scale *is logarithmic* and as a result, each whole pH value below 7 is ten times more acidic than the next higher value. For example, pH 4 is *ten times more acidic* than pH 5 and 100 times (10 times 10) more acidic than pH 6. The same holds true for pH values above 7, *each of which is ten times more alkaline* (another term for *basic*) than the next lower whole value. For example, pH 10 is ten times more alkaline than pH 9 and 100 times (10 times 10) more alkaline than pH 8.[157]

And

> Over the past 300 million years, ocean pH has been slightly basic, averaging about 8.2. Today, it is around 8.1, a drop of 0.1 pH units…[158]

What does this testimony tell the jury?

Acidification refers to the *direction* of pH change. Lowering *any pH* is considered acidification *regardless of the actual pH of the solution* (the solution might be alkaline, neutral, or acidic).

Alkalization also refers to the direction of pH change. *Raising any pH* is considered alkalization regardless of the actual pH of the solution (the solution might be alkaline, neutral, or acidic). The most common reference to any lowering of pH is acidification regardless of the pH value (acidic or alkaline).

While ocean water pH may vary considerably in highly localized areas (it may even be acidic), as noted above the vast bulk of ocean water typically has an alkaline pH averaging about 8.1.

Either atmospheric CO_2 growth or colder climate could lead to relatively greater absorption of CO_2 from the atmosphere *making oceans less alkaline. That process is known as ocean acidification.*

Yet ocean water *is not* acidic, it remains firmly alkaline (basic). Even though it is alkaline, cold ocean absorption of excess CO_2 often leads to a chemical process that creates an acid with the result that ocean pH can be slightly lowered. That is the basis upon which the claim of ocean acidification rests.

Suppose the prosecution, instead of claiming "ocean acidification must be caused by anthropogenic carbon dioxide (ACO_2) emissions," had claimed "ocean de-alkalization must be caused by ACO_2 emissions." How different would the public view the alternate statement describing exactly the same condition?

The term *acidification* has a scary implication and the prosecution uses that for its propaganda value. Such claims were ramped up during the "pause" in meaningful climate warming that was notorious since the early 2000s. The prosecution had to maintain a constant reminder that fossil fuel use is evil and will destroy the planet. So ocean acidification became a suitable ally of the human-caused global warming mythology.

Because oceans have a natural buffering against becoming acidic, it is misleading for the prosecution to lead jurors to believe atmospheric CO_2 growth will ultimately create acidic oceans. It will not. As Prof. Tom V. Segalstad explains in "Some Thoughts on Ocean Chemistry"[36] (*Climate Change Reconsidered II: Biological Impacts*, Chapter 6.3.1.2):

> It is important to note the dissolution of CO_2 in water is governed by Henry's Law, evidenced by the fact there is approximately 50 times more CO_2 dissolved in the ocean than in the atmosphere at present. It is this vast mass of dissolved CO_2 in the ocean that holds the regulating power—not the relatively small amount of CO_2 contained in the air. Furthermore, the chemical reaction speeds involved in the dissolution of CO_2 are high, as is the ocean circulation speed in the upper parts of the ocean.

> The ocean acidification hypothesis also ignores the presence of vast amounts of dissolved calcium in the ocean: the upper 200m of ocean water contains enough dissolved calcium to bind all anthropogenic CO_2 as precipitated calcium carbonate (in the ocean) without affecting the ocean's pH

(Jaworowski et al., 1992a; Segalstad, 1996; 1998). The ocean acidification hypothesis also ignores or downplays other oceanic buffers (pH stabilizing reactions), the thermodynamic stability of solid calcium carbonate in ocean water, and photosynthesis by marine biological systems.

Oceans are not acidic. Lowering ocean pH from 8.2 to 8.1 is acidification because oceans were alkaline at pH of 8.2, and they have become slightly less alkaline at pH of 8.1, yet at no time have oceans ever been characteristically acidic. Oceans have been in the past, are presently, and most likely always will be alkaline.

Beware of dire warnings about ocean acidification because such claims may well be intended to deliberately deceive and frighten jurors.

In a 2016 article,[159] "Ocean Acidification: Yet Another Wobbly Pillar of Climate Alarmism," James Delingpole reviewed the claim and found "many studies are flawed, and the effect may not be negative even if it's real."

In that article, Delingpole reports:

> Ocean acidification is the terrifying threat whereby all that man-made CO_2 we've been pumping into the atmosphere may react with the sea to form a sort of giant acid bath. First it will kill off all the calcified marine life, such as shellfish, corals and plankton. Then it will destroy all the species that depend on it—causing an almighty mass extinction which will wipe out the fishing industry and turn our oceans into a barren zone of death.

The reality may be rather more prosaic. Ocean acidification—the evidence increasingly suggests—is a trivial, misleadingly named, and not remotely worrying phenomenon… According to Patrick Moore, a co-founder of Greenpeace,… the term is "just short of propaganda." The pH of the world's oceans ranges…well above the acid zone…more correctly it should be stated that the seas are becoming slightly less alkaline. "Acid" was chosen, Moore believes, because it has "strong negative connotations for most people."

Matt Ridley…wrote, "Ocean acidification looks suspiciously like a back-up plan…in case the climate fails to warm.'"…That's why I like to call it the alarmists' Siegfried Line—their last redoubt should it prove, as looks increasingly to be the case, that the man-made global warming theory is a busted flush.

To the alarmist camp…"deniers" are heartless, anti-scientific conspiracy theorists who don't read peer-reviewed papers…Unfortunately for the doom-mongers, we sceptics have just received some heavy fire-support from a neutral authority.

Howard Browman, a marine scientist for 35 years, has published a review…of all the papers published on the subject. His verdict could hardly be more damning. The methodology used by the studies was often flawed; contrary studies suggesting that ocean acidification wasn't a threat…had difficulty finding a publisher.

Ocean acidification theory appears to have been fatally flawed almost from the start. In 2004… NOAA scientists…produced a chart showing a strong correlation between rising atmospheric CO_2 levels and falling oceanic pH levels. But then, just over a year ago, Mike Wallace, a hydrologist with 30 years' experience, noticed while researching his PhD that they had omitted some key information. Their chart only started in 1988 but, as Wallace knew, there were records dating back to at least 100 years before. So why had they ignored the real-world evidence in favour of computer-modelled projections?

When Wallace plotted a chart of his own, incorporating all the available data, covering the period from 1910 to the present, his results were surprising: there has been no reduction in oceanic pH levels in the last century.

Even if the oceans were "acidifying"…marine species that calcify have survived through millions of years when CO_2 was at much higher levels; second, they are…capable of adapting…to environmental change; third, seawater has a large buffering capacity which prevents dramatic shifts in pH; fourth, if oceans do become warmer due to 'climate change', the effect will be for them to "outgas" CO_2, not absorb more of it.

Finally, and perhaps most damningly, Moore quotes a killer analysis conducted by Craig Idso of all the studies which have been done on the effects of reduced pH levels on marine life. The impact on calcification, metabolism, growth, fertility and survival of calcifying marine species when pH is lowered up to 0.3 units (beyond what is considered a plausible reduction this century) is beneficial, not damaging.

This raises several key points for the jury to consider:

- First, the term *ocean acidification*, while technically correct, is nevertheless inappropriate as it may mislead the public. The phrase *less alkaline* is equally true, means the same thing, and does not run the risk of misleading those not familiar with pH (few jurors are chemists or chemical engineers).
- Second, ocean pH change is not a straightforward issue and clearly is neither as critical nor as certain as the prosecution would lead the jury to believe.
- Third, given the lack of adequate records to gain a meaningful perspective on the historic variability of ocean pH during periods when atmospheric CO_2 was much greater than it is today and/or climate was much colder, the degree of ocean alkalinity and recent alkalinity change cannot reasonably be characterized as alarming.
- Finally, there is no evidence oceans have ever been acidic even during periods of vastly higher levels of atmospheric CO_2 and much colder climate.

Summary: Ocean Acidification

1. The term *ocean acidification*, while chemically correct, runs the risk of being deceptive. A clearer alternative, *reduced ocean alkalinity*, is a less deceptive term for the same condition.
2. Oceans remain *firmly alkaline* (at pH = 8.1 > neutral 7).
3. Minor variations in ocean alkalinity *do not pose a threat* to either ocean life or life in the global biosphere.
4. There is *insufficient evidence* to link changes in anthropogenic CO_2 to ocean pH variability.

5. Much higher historic levels of atmospheric CO_2 (averaging over 2100 ppm for the past 550 million years) *have never caused mass extinctions* or turned oceans into acidic reservoirs devoid of life. There is *ample evidence that ocean life thrived* during periods of high atmospheric CO_2.
6. Beyond a trivial impact that does not rise to the level of an indictable offense, there is insufficient evidence to find carbon dioxide guilty of this prosecution allegation.
7. Oceans contain roughly 50 times as much CO_2 as the atmosphere. It is implausible that relatively small changes in atmospheric CO_2 would have a significant impact on ocean CO_2 content and pH.
8. Are anthropogenic CO_2 (ACO_2) emissions guilty of causing significant ocean acidification? No. There is insufficient ACO_2 to be the source of any detectable shift in ocean pH.

Cow Flatulence and Atmospheric Methane

Are claims that cow flatulence has a meaningful impact on climate change serious?

Evident in the largest magnification in Figure 3-1 (page 29), the atmospheric volume of methane (CH_4) is about 2 ppm, a tiny fraction of CO_2's atmospheric volume (~408 ppm), meaning 99.9998% of atmospheric gases *are not* CH_4.

To put into perspective just how little of the atmosphere contains CH_4, recall the analogy from Chapter 3 that relates the Rose Bowl stadium's 92,542 seats to the amount of carbon dioxide in the atmosphere. Methane is so rare in the atmosphere that if seven Rose Bowl stadiums with 647,794 seats represented all atmospheric gases, *less than two (1.3) of those seats would represent all atmospheric methane.* This underscores the audacity of the prosecution's claim that methane is capable of noticeably altering global climate!

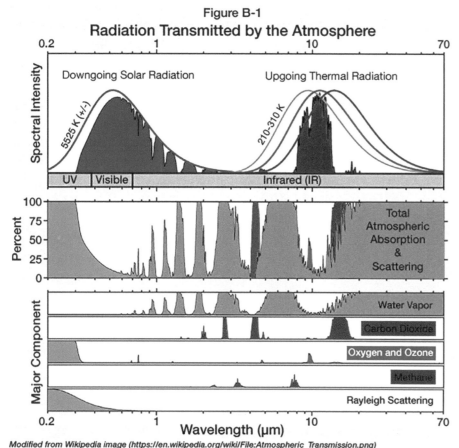

Figure B-1
Radiation Transmitted by the Atmosphere

Modified from Wikipedia image (https://en.wikipedia.org/wiki/File:Atmospheric_Transmission.png)

The prosecution routinely deceives the jury by focusing on carbon dioxide (CO_2) and methane (CH_4) being highly reactive to radiant heat (IR). Yet the prosecution ignores exculpatory evidence by ignoring (1) their scarcity in the atmosphere and (2) the extremely limited spectra of the IR to which they *are* reactive (Figure B-1).

The bottom panel of Figure B-1 shows the limited wavelengths of radiant heat (IR) to which both CO_2 (brown) and CH_4 (yellow-green) are reactive. Remember, both are trace gases (constituting much less than 1% of atmospheric gases). Both have *very limited* ranges of the IR to which they are reactive (most of the IR spectrum is nonreactive with CO_2 and CH_4). Note, particularly, the extremely limited range of the IR spectrum reactive with methane. CH_4 is invisible to all other wavelengths of IR and will have no reaction to IR radiated from Earth at those wavelengths.

The middle panel shows the atmosphere's absorption of IR by all the indicated greenhouse gases. Note that water vapor (light blue) dominates the absorption of IR over a very large range of wavelengths. That, coupled with water vapor averaging 1.5% of atmospheric gases makes water vapor, by far, the most potent greenhouse gas.

The limited impact of other greenhouse gases is shown by comparing the lower panel with the middle panel. That comparison shows that the two lowest wavelengths where CO_2 is reactive to IR contribute nothing as they are outside the range of significant radiant IR given off by Earth.

The next higher wavelengths where IR is reactive with atmospheric CO_2 is over a very limited range of Earth's outbound IR (4 to 5 μm) so that outbound IR at those wavelengths would have very limited impact on global climate. The largest range of CO_2 reactivity to IR (at wavelengths between 13 and 18 μm, middle panel) also has limited impact because water vapor is a strong competitor for the atmospheric absorption of IR in that region.

The upper panel illustrates the wavelengths of incoming solar radiation (invisible ultraviolet, visible light, invisible infrared) and the range over which Earth's outbound radiant heat (IR) occurs (bell-shaped curves). The dark blue area indicates wavelengths where solar radiation penetrates the atmosphere. The red area is where few atmospheric gases "see" (are reactive to) outbound IR allowing radiant heat at those wavelengths to pass through the atmosphere virtually unimpeded.

Inescapable conclusions the jury can draw from Figure B-1 is (1) the power of greenhouse gases to react with Earth's outbound IR is dominated by water vapor and (2) as the evidence shows, the prosecution vastly overstates the potential impact of both CO_2 and CH_4 on atmospheric temperature.

There is a third devastating reality that demonstrates fossil fuel emissions have little capacity to impact climate change. This additional knife in the greenhouse gas theory might be termed "the law of diminishing returns".

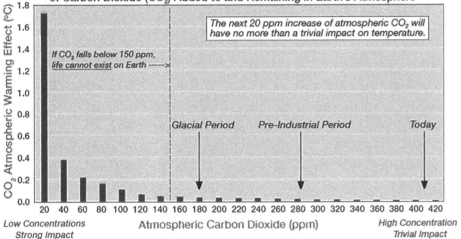

Figure B-2
Dramatic Reduction of Potential Warming from *Each Additional 20 ppm* of Carbon Dioxide (CO_2) Added to and Remaining in Earth's Atmosphere

Source: Based on "The Logarithmic Effect of Carbon Dioxide" by David Archibald published at https://wattsupwiththat.com/2010/03/08/the-logarithmic-effect-of-carbon-dioxide/

Figure B-2 illustrates why the jury hasn't heard the prosecution mention this detail to the jury.

At the current level of atmospheric CO_2 another 20 ppm (from, 410 ppm to 430 ppm) has the potential to warm the atmosphere by about 0.01°C.

The notion that simply pouring more CO_2 into the atmosphere will have a sudden dramatic impact on global climate is not supported by either the evidence or sound science.

The science is clear and it testifies that at current levels atmospheric CO_2 is neither a "strong" climate change force, nor is there any capacity for future emissions of fossil fuel CO_2 to have any noticeable impact on global average surface temperature (climate change).

This diminishing effect, combined with the two problems cited previously, helps to explain why current and future levels of greenhouse gas growth cannot have a measurable impact on climate change. It also explains why the 550-million-year global average atmospheric CO_2 of over 2100 ppm did not create catastrophic warming. Indeed, some of the highest levels of atmospheric CO_2 (more than 10 times the current level) accompanied some of the coldest climate in Earth's history (about 450 million years ago).

Diminishing returns combined with the fact that CO_2 and CH_4 are trace gases in the atmosphere, both having *very limited* wavelengths of the IR over which their molecules react to IR is a fatal blow to the over-stated claims made by the prosecution throughout this trial. Supporting this evidence is the mountain of testimony given in Chapters 3–5 that confirms no detectable relationship exists between growing atmospheric CO_2 and changing climate.

Over a one-hundred-year period, methane has been claimed to be twenty-eight times more potent as a greenhouse gas than is carbon dioxide.[160] Since the records clearly show atmospheric CO_2 has *no discernible relationship with climate change* beyond pure chance, in terms of methane's impact on climate change, twenty-eight times zero is still zero!

Believing it has fooled the jury into accepting its claims that carbon dioxide is a potent climate change force, the prosecution attacks sources of dairy and meat as having a significant impact on global climate because dairy animals pass methane into the atmosphere as part of their digestive process. Such claims aren't science, they're lunacy.

The prosecution tells the jury that methane is even more highly reactive than carbon dioxide. In the laboratory where the wavelength of IR can be tightly controlled, that is true. But what the prosecution *doesn't tell the jury* is that the amount of methane in the atmosphere from all sources is 1/200th (0.005) the amount of carbon dioxide in the atmosphere and that, while CH_4 is more reactive than CO_2, it is reactive over just two *very small portions* of the full spectrum of infrared radiation (Figure B-1). Water vapor's atmospheric absorption of outbound IR dominates and overwhelms any warming effect of CO_2 and CH_4.

Summary: Cow Flatulence and Atmospheric Methane

Even if CH_4 were capable of exercising its full potential to react to outbound IR, several key factors prevent methane gas from trapping detectable heat in the atmosphere:

- Little of the IR spectrum reacts with either CO_2 or CH_4 molecules. CH_4 is reactive over a particularly small portion of the IR spectrum being radiated from Earth. Most outbound IR is unaffected by either CO_2 or CH_4.
- CH_4 is so rare in the atmosphere; it has very limited capacity to impact atmospheric temperature.
- The few wavelengths of the IR spectrum reactive to CH_4 are already dominated by water vapor's IR absorption.
- There is no evidence in any record of changing atmospheric CH_4 that testifies to such change having any detectable impact on either global temperature or climate.
- Atmospheric methane is not a climate change *force*, it is a climate change *farce*.

More About USCRN Records (2005–2019)

As mentioned earlier in Chapters 4 and 5, the USCRN's 114 new temperature measuring stations distributed across the USA are providing the highest quality temperature records, free of urban heat island effects and other siting problems that impact the quality of legacy temperature measurements.

Figures B-3 and B-4 (next page) show 14-year (2005-2019) net temperature change for each month and season.[174] Net temperature change between any two years is simply the difference between the anomalies for each of those years. For example, the net temperature change between 2005 and 2019 for the month of January is calculated by subtracting the January 2005 anomaly from the January 2019 anomaly.

These USCRN records offer jurors compelling testimony that contradicts the prosecution's climate change theory.

Figure B-3 (14-year monthly aggregate temperature change) shows the (2005–2019) temperature change for each month. The seven months during which temperature cooled over the 14-year period of change are shown in blue; the five months with 14-year warming are shown in red. Overall, cooling dominates the 14-year change with the net change averaging -0.77°F (-0.43°C) and monthly weighted (by days in the month) net change averaging -0.74°F (-0.41°C).

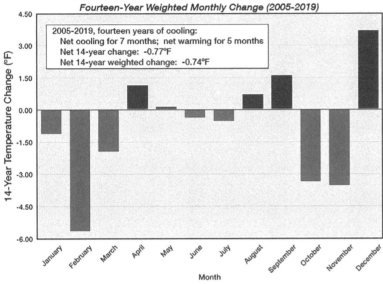

Figure B-3
USCRN 114-Station Continental US Records
Fourteen-Year Weighted Monthly Change (2005-2019)

2005-2019, fourteen years of cooling:
Net cooling for 7 months; net warming for 5 months
Net 14-year change: -0.77°F
Net 14-year weighted change: -0.74°F

Source: USCRN 114-station continental US average monthly temperature anomalies (2005-2019)

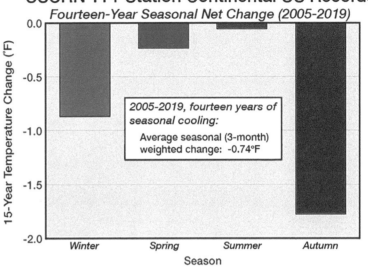

Figure B-4
USCRN 114-Station Continental US Records
Fourteen-Year Seasonal Net Change (2005-2019)

2005-2019, fourteen years of seasonal cooling:
Average seasonal (3-month) weighted change: -0.74°F

Source: USCRN 114-station continental US average monthly temperature anomalies (2005-2019)

Figure B-4 shows the seasonal (weighted by days in the season) net temperature change over the same 14-year period of change. Note that every season registered a net cooling over the fourteen years of change. Table B-1 (next page) summarizes the monthly and seasonal 14-year net temperature change.

Table B-2 (next page) shows how 14-year monthly temperature change is calculated from the difference between the anomaly for each of the two years over which temperature change is being calculated. Yearly anomaly is the sum of the weighted monthly anomalies for that year. Net temperature between the two years is the sum of the weighted monthly differences for each year. Table B-2 shows how the net 14-year cooling of -0.74°F (-0.41°C) is calculated for the 14-year period between 2005 and 2019.

Table B-3 (next page) lists average yearly change for both MLO CO_2 (ppm per year) and USCRN temperatures (°F per year) together with the correlation coefficient between yearly CO_2 and temperature changes. A correlation coefficient of -0.0839 means annual CO_2 and USCRN temperature changes are highly uncorrelated, and, in fact, the negative sign indicates as CO_2 increases, temperature decreases!

Clearly, there can be no causal relationship between the two. In addition, the average yearly change of MLO atmospheric CO_2 continues to grow (at 2.26 ppm per year) while the yearly change of USCRN temperature shows an average cooling at the rate of -0.05°F (-0.03°C) per year! Persistently growing atmospheric CO_2 while USCRN temperatures cool is further compelling testimony against the prosecution's theory of a crime.

Table B-1
Fourteen-Year Monthly & Seasonal* Net Change
USCRN 114-Station Continental US Records (2005-2019)

Month	14-year Monthly Change (°F)	14-year Seasonal *Weighted Change (°F)
January	-1.12	-0.87
February	-5.65	
March	-1.95	-0.24
April	1.14	
May	0.13	
June	-0.37	-0.06
July	-0.54	
August	0.72	
September	1.60	-1.78
October	-3.36	
November	-3.53	
December	3.69	
Net 14-year weighted change	-0.74°F	

Source: USCRN 114-station continental US average monthly temperature anomalies (2005-2019)

Winter ▢ Spring ▢ Summer ▢ Autumn ▢

Table B-2
Calculating USCRN 2005-2019 Temperature Change
14-Year Change = Anomaly (2019) - Anomaly (2005)
Yearly Anomaly = Monthly Weighted Sum of Anomalies

Month	2005	2019	14-Year Change
January	1.75	0.63	-1.12
February	2.50	-3.15	-5.65
March	-0.88	-2.83	-1.95
April	0.41	1.55	1.14
May	-1.26	-1.13	0.13
June	0.23	-0.14	-0.37
July	1.35	0.81	-0.54
August	0.45	1.17	0.72
September	2.05	3.65	1.60
October	1.13	-2.23	-3.36
November	2.11	-1.42	-3.53
December	-0.50	3.19	3.69
Weighted Yearly Anomaly	0.76°F	0.02°F	-0.74°F

Every 14-year weighted seasonal (3-month) change shows cooling. The greatest cooling is in autumn, -1.78°F (-0.99°C), and winter, -0.87°F (-0.48°C); the least cooling in summer, -0.06°F (-0.03°C); and spring has modest cooling. -0.24°F (-0.13°C). From 2005 to 2019 the overall monthly-weighted temperature change of the 114 USCRN monitoring stations is -0.74°F (-0.41°C) of cooling.

Jurors are cautioned to bear in mind that these stations measure surface temperatures over land which is more responsive to temperature change forces. Furthermore, the land monitored is confined to the continental USA. Nevertheless, the USCRN records, arguably the most accurate anywhere on the planet, fly in the face of the prosecution's claims that recent years have experienced "unprecedented warming" and are among "the warmest on record" when that is clearly not the case. Indeed, as Figure B-5 illustrates, the USCRN records show a dominance of cooling over the fifteen years during which the USCRN reporting stations have been active.

Table B-3
2005-2019 Average Yearly Change
MLO CO₂ & USCRN Temperatures

Correlation Coefficient	-0.0839
Mauna Loa Observatory CO₂ Change	+2.26 ppm/year
USCRN Weighted Temperature Change	-0.05°F/year

Figure B-5 shows the change through 2019 for each year beginning with 2005 and notes that of the one to fourteen years of change, ten experienced cooling to 2019 with only four warming to 2019—clear evidence that shows the year 2019 was not among the warmest on record.

With varied terrain from flat plains to rugged young mountains across both temperate and subtropical climate zones, including both maritime coastal and interior desert lands far from major bodies of water, these 114 monitoring stations have broad coverage of a wide variety of land surface types.

If persistently growing atmospheric CO_2 were truly a "strong climate change force," then a temperature change of -0.74°F in the USCRN records over the past fourteen years of change would be highly unlikely and would demand an explanation. But the USCRN record, the most accurate land surface temperature monitoring network on the planet, is not referenced in current news reports and is virtually ignored by the prosecution.

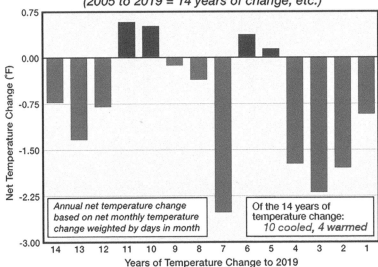

Figure B-5
USCRN Net Temperature Change to 2019
(2005 to 2019 = 14 years of change, etc.)

Source: USCRN (NOAA) Contiguous U.S. Monthly Average Temperature Anomaly (2005-2019)

Defying the prosecution's theory of a climate change crime, while atmospheric CO_2 continued its persistent climb (~30 ppm) between 2005 and 2019, the USCRN 114-station temperatures reported a net temperature cooling of -0.74°F (-0.41°C)! These records confirm there isn't a scintilla of any theory-consistent relationship between persistently growing atmospheric CO_2 and USCRN temperatures.

This powerful evidence is untainted by either unreliable legacy temperature monitoring stations or the dubious temperature "homogenization" activities ostensibly designed to correct tainted temperature records.

At some point, the obvious question needs to be asked. What would it take to get the prosecution to drop its doomed case against atmospheric CO_2 in general and fossil fuel emissions in particular? If compelling evidence is insufficient, what *would be* sufficient?

Other Climate Change Forces

Solar Activity

The jury learned[46–48] that the Sun's true average (mean) irradiance is not precisely known but is estimated to be somewhere in the range of 1360 to 1365 watts per square meter (Wm^{-2}) with an estimation error of 5 Wm^{-2}.

According to the prosecution's theory-based testimony (IPCC's 2013 *Fifth Assessment Report*), the estimated entire human impact on climate since 1750 amounts to a mere 2.3 Wm^{-2} or *less than half* the range of uncertainty (estimation error) associated with measuring solar irradiance![47]

This means the prosecution really has no idea how much of measured temperature variability is due to its perceived human impact and how much is a consequence of its inability to precisely measure solar irradiance! So the prosecution's estimate of the human impact on climate change is really nothing more than theory-based speculation. Jurors must decide on the basis of evidence, not speculation.

The magnitudes of solar grand maxima and minima are known to have a high correlation with global climate change, the former being associated with the Medieval Warm Period (MWP) and the latter being associated with the Little Ice Age (LIA). As the recent late 20th-century to 21st-century solar grand maxima approached, mid-20th-century climate cooling ceased (Figure 3-4, page 43), and post-Little Ice Age[11] slight warming resumed.

Geologic Activity

El Niño Southern Oscillation (ENSO) cycles are:

> An irregularly periodic variation in winds and sea surface temperatures over the tropical eastern Pacific Ocean, affecting climate of much of the tropics and subtropics. The warming phase of the sea temperature is known as El Niño and the cooling phase as La Niña.[121]

ENSO oscillations (between El Niño, neutral, and La Niña) are known to affect global circulation patterns and climate. Evidence of ENSO cycles go back to at least 1876, lending credibility to a relatively new theory that suggests geyser cycles are analogous to geologic forces that trigger ENSO cycles.[24, 52] It is believed porous layers along the sides of the Mariana Trench periodically become saturated with superheated water. The superheated water is subsequently expelled in a grand eruption of very warm water that rises to the surface of the western Pacific Ocean and migrates over the Pacific Ocean surface to the east coast of South America. These cycles can have a profound effect on short-term warming and cooling patterns and may trigger other forces whose impact is longer term.

Volcanic eruptions such as the massive 1815 Tambora and 1883 Krakatoa eruptions have measurably cooled global climate. The year 1816 is known as "the Year Without a Summer" in New England; six inches of snow fell in June and every month of that year recorded a hard frost (a "hard frost" generally occurs when temperatures fall below 28° Fahrenheit for a few hours).[10]

There is evidence of active volcanic eruptions under the Arctic Sea that are believed to be playing a role in Arctic Sea ice reduction.[161] It is suggested that melting is enhanced because more eruptions trigger greater melting that leads to more eruptions. The rationale for this feedback cycle is that sea ice "melt lighten[s] the load on the crust of the earth, which lets the magma underneath rise."[161] Because sea ice weighs the same as seawater, any additional sea ice weight is from the accumulated snowfall and freshwater ice on top of the sea ice.

Summary: Other Climate Change Sources

Well-known forces that cause both short-term (several years) and long-term (decades to hundreds of years) climate cycles are good candidates for some of the climate change observed since the end of the LIA. None of these climate variations are either unusual or extreme in Earth's climate history.

Why the Deceptions?

Listed below are familiar claims offered by the prosecution and its agents in their continuing effort to mislead the jury and convince the populace that, despite a mountain of evidence to the contrary, its human-induced climate change theory is valid and cause for alarm. Following each claim is a summary of Defense's refutation of that claim:

The Science Is Settled

Scientific laws are settled. Theories and hypotheses are *never* settled. Competent, honest scientists would *never* claim a theory is settled science.

Einstein's theory of relativity is not settled science (if it were settled, it would be Einstein's *law* of relativity). Einstein's theories of general and special relativity are to this day being tested and challenged in a process that is not antagonistic but which seeks to discover any flaws or special cases that require the theory to be modified.

That is how *real* science advances; it doesn't advance by fiat, consensus, or a dogmatic defense of theory.

Any proposition based on a theory that is contradicted by the clear testimony of real-world evidence is not "settled" science. Those who claim otherwise are either ignorant or charlatans. In either case, their word cannot be trusted.

If the science were really as solid as the prosecution claims, why go to such lengths to deceive the jury? Most important, why do real-world observations unequivocally contradict the prosecution's *climate change* theory?

A 97% Consensus of Scientists Agree...

Real science is never determined by consensus. Questions of science are best resolved by a process used for centuries and known as *the scientific method* (defined in the glossary).

Every theory (or hypothesis) should undergo careful scrutiny using *the scientific method*, a process to rationally and logically examine theory and hypothesis for compliance with (1) known laws of science, (2) observation in nature, and/or (3) experimentation. Such examination expands knowledge by either supporting the proposition, rejecting the proposition as invalid, or indicating the proposition is defective and requires modification to satisfy its observed defects.

The essence of *the scientific method* encourages scientists to subject *every* theory to an objective skeptical scrutiny. Uniformity of thought, prejudice, avoidance of debate, dogmatic rejection of contrary views and denial of nonconforming observation and experimentation are entirely inconsistent with sound scientific methods. Yet such actions have characterized the prosecution's dogmatic defense of its theory.

The prosecution's agents often claim "97% of scientists agree…"—a figure based on Australian professor John Cook's subjective review of 11,000 *abstracts* of scientific papers relating to climate change. An abstract of Cook's paper noted that of the 11,000 papers, 66.4% offered *no position* on anthropogenic global warming (AGW) while *in Cook's opinion* only 32.6% endorsed AGW with *97.1% of that 32.6%* agreeing with the prosecution's climate change theory (31.65% of all abstracts). Again, these figures are based on assessments of the abstracts surveyed, not on a reading of the actual studies. So in fact, of the 11,000 papers, *at most* 3,482 endorsed the prosecution's climate change theory, *whereas 7,518 did not endorse it.*

Evidently, this is an example of the prosecution's new math where 97% = 32%. With that precision, it is easy to understand how the prosecution's scientists cannot draw an informed conclusion from a review of the evidence.

Imagine if, instead of "97% of…" the jury heard "Just 32% of…" or "A 68% majority of scientists *disagree*…"!

Skeptics Are "Deniers" of Climate Change

A form of name-calling devised to marginalize those who scrutinize the prosecution's climate change theory and find serious weaknesses in the prosecution's case. As the testimony of Earth's climate history supports, those who believe recent climate change is perfectly natural and well within typical historic bounds are pejoratively branded "deniers" or "skeptics" of the prosecution's flawed theory. References to skeptics

and deniers are generally made in a dismissive pejorative manner specifically designed to avoid any serious discussion of the facts.

Contrary to prosecution allegations, skeptics of the prosecution's dubious theory do not deny climate changes; they deny humans have any discernible impact on climate change, a position fully supported by real-world evidence.

Such name-calling is an awkward attempt to appeal to authority by posturing the prosecution as authority and implying those who question the prosecution's theory lack the authority to do so. Appeals to authority are the resort of those who believe the prosecution but who are ill-equipped to argue the prosecution's case. Which raises the question, on what basis is their belief rooted?

Quotable:

> The improver of natural knowledge absolutely refuses to acknowledge authority, as such. For him, skepticism is the highest of duties; blind faith the one unpardonable sin.
>
> —(T. H.) Huxley

> When the debate is lost, slander becomes the tool of the loser.
>
> —Socrates

"Climate Change" as Shorthand for "Human-Induced Climate Change"

The term global warming was used to describe natural and unremarkable multidecadal warming circa 1980. In the 1990s, the term became associated with the assumption that human activity, principally fossil fuel combustion, is responsible for warming on the basis of an untested theory that so-called greenhouse gases (primarily water vapor and carbon dioxide) were significant forces affecting climate change (in this case, warming).

As the chasm between model-projected climate and actual observations broadened (Figure 5-7, page 101) with climate stubbornly refusing to warm at the extreme rates projected by climate change theory, it became too embarrassing for the prosecution to continue talking about global warming when little warming was occurring, so "climate change" became the favored new term.

Since climate is always changing (cooling, warming, or on the cusp of change), it is deceptive to co-opt the phrase climate change (which is real) as a substitute for human-induced climate change, which is theoretical, unproven, dubious, and cannot be measured.

The use of the term global warming continued into the 21st century despite a hiatus (or "pause") in the rate of warming after 1998 that has persisted for more than two decades. As the solar grand maximum fades into the past, the hiatus is continually being confirmed by evidence gathered from satellite and weather balloon (radiosonde) observations that, with the exception of two observed strong El Niño deviations, indicate late 20th- and early 21st-century temperatures have become relatively stable with only a trace of the long-term natural warming trend observed following the end of Little Ice Age cooling in the mid-19th century. Early 21st century (2005 to 2019) USCRN-observed cooling may portend the end of post-LIA warming.

The subsequent shortening of "human-induced climate change" to just "climate change" served to confuse the jury by implying that all climate change is "human induced"—yet another deception.

Homogenized Temperature Records

Official historic surface station temperature records in widespread areas of several nations have been deliberately adjusted (tampered with). Homogenization's[91–94, 97, 99–102] effect is that an overwhelming preponderance of records now show recent cooling years have been warmed and past warming years have been cooled. These changes cause temperature histories to create an impression of long-term steep warming that is not seen in either the untampered ground station records or the satellite and radiosonde records.

Clearly designed to support the climate change narrative, these changes have created out of thin air a grossly distorted view of climate history.

A good example of this brazen evidence-tampering is the homogenizing of rural temperature records that were considered accurate to make them consistent with warmer nearby urban temperature records known to be inaccurately high due to the urban heat island effect. Rather than justifiably cooling the artificially high urban temperatures, the accurate rural temperatures in surrounding areas were *raised* to be consistent with the inaccurate elevated urban temperatures, expediently creating evidence of (human-induced) global warming! Elevated suburban-rural temperatures were created by humans tampering with raw data![94] Homogenization definitely qualifies as "man-made" evidence of climate change.

Over a period of years, these adjustments to the raw temperature records have succeeded in artificially appearing to have eliminated the dust bowl heat of the 1930s that had (until homogenization) been acknowledged to be the hottest decade ever recorded in the USA.

There is no legitimate need to be adjusting surface station raw temperature records by homogenizing them. Clearly, this is just a case of the prosecution and its agents tampering with evidence in a desperate effort to try to create a fake temperature record that fits its concocted theory to further deceive the jury.

Assumption-Based "Certainty"

The Global Carbon Project is an agent of the prosecution and relies on the pronouncements of the prosecution for its testimony. This is the equivalent of the prosecution testifying for the prosecution! On its website under the heading "Global Carbon Budget," it repeats the prosecution's narrative based on its assumed atmospheric lifetime for anthropogenic CO_2 that claims it to be 10 to 40 times higher than determined by traditional highly-competent scientific experimentation performed by chemists and chemical engineers over many decades. This assumption-based approach makes a mockery of science that leads to assertions that are flat-out incorrect because they are rooted in a flawed assumption that anthropogenic emissions *must be* responsible for atmospheric CO_2 growth.

The following propaganda emerges:[162]

> Of the total emissions from human activities during the period 2007–2016, about 46% accumulated in the atmosphere, 24% in the ocean and 30% on land. During this period, the size of the natural sinks grew in response to the increasing emissions, though year-to-year variability of that growth is large. The strength of the 2016 ocean CO_2 sink was above the decadal average and the land sink below average. Both trends are consistent with a positive phase of El Niño. The total estimated sources do not match the total estimated sinks, creating the carbon imbalance. This imbalance reflects the gap in our understanding and results from the uncertainties from all budget components.

In a mockery of science, the prosecution devised a scheme to force the atmospheric accumulation of anthropogenic CO_2 to match the observed growth of atmospheric CO_2 by manufacturing an unrealistic extremely long atmospheric lifetime for fossil fuel emissions.

The percentages cited in the quoted paragraph (under a heading "CO_2 Removals by Natural Sinks") are derived from the manufactured evidence based on the science-defying claim that unremarkable emissions of CO_2 from the combustion of fossil fuel resist reabsorption for 10 to 40 or more times longer than all other identical plant-derived CO_2 emissions![31]

The actual percentage of "human activity" emissions that accumulate in the atmosphere is dramatically lower than the claimed 46%. Using realistically-determined emissions' lifetimes of 4 to 6 years, atmospheric accumulation ranges from 3% to 6%.[31–32, 139]

Circular speculation is a common prosecution practice, using one assumption to support yet another assumption and then wrapping the entire package of speculative assumptions in the aura of sound scientific evidence.

The real travesty is that the prosecution's handiwork then appears online and in publications designed to create the illusion of sound scientific testimony when it is indistinguishable from rank speculation presented as scientific certainty.

Summary: Deceptions

Trying to sway the jury into believing its theory of a crime, the prosecution engages in a host of deceptions designed to steer the jury to an improper verdict.

These deceptions include but are not limited to:

- *The science is settled:* Theory *is never* settled; only scientific laws are considered settled.
- *A 97% consensus of scientists agree:* Science *is never* determined by consensus. The 97% figure is fiction.
- *Skeptics are deniers of climate change:* The essence of science is skeptical scrutiny of all hypotheses, theories, and propositions. Pejorative name-calling is bullying designed to deflect serious scrutiny of greenhouse gas climate change theory. When honest debate fails, charlatans resort to name-calling.
- *Climate change as shorthand for human-induced climate change:* A propaganda tactic designed to train the jury to automatically link natural climate change to human activity, suggesting climate never changes naturally. Originally used to avoid the embarrassment of referring to "global warming" during the well-documented non-ENSO-related warming pause during the first two decades of the 21st century.
- *Homogenized temperature records:* If the record doesn't match the theory, change the record. Manipulating the record stands *the scientific method* on its head and further calls into question the prosecution's methods.
- *Assumption-based "certainty":* The prosecution laces its allegations that underpin its indictment of CO_2 with assumption-based absolutes (e.g., "must be") to deceive the jury.

For more about the scope of IPCC (prosecution) inconsistencies and deceptions, read Dr. Jeffrey Glassman's paper, *CO₂: 'Why Me?'—On Why CO_2 is Known NOT to Have Accumulated in the Atmosphere & What is Happening with CO_2 in the Modern Era.*[33]

A Final Example of How Wrong Prosecution Agents Can Be

Speaking to an audience in 2008, assistant prosecutor Al Gore notoriously predicted that "the entire north polar ice cap during some of the summer months could be completely ice-free within the next five to seven years." Did anyone notice an ice-free Arctic in any of the summers from 2013 to 2015? How about since 2015? Was the Arctic Sea *anything close* to being ice-free? No, it wasn't. Once again, that particular prosecution agent was just plain inconveniently (and spectacularly) wrong.

Not content to be dead wrong about polar ice, assistant prosecutor Gore famously predicted a 20-foot sea level rise. He conveniently failed to provide a specific timeframe. In late 2016, an article documented four different studies that found *"no observable sea-level effect of anthropogenic global warming."*[163]

On February 7, 2014, an article, "The End of Snow?"[164] by Porter Fox in *The New York Times* stated:

> Daniel Scott, a professor of global change and tourism at the University of Waterloo in Ontario, [analyzed] potential venues for future Winter Games. [He cautioned that] by 2100, there might not be that many snowy regions left in which to hold the Games. He concluded that of the 19 cities that have hosted the Winter Olympics, as few as 10 might be cold enough by midcentury to host them again.

Less than a month later, Monday, March 3, 2014:

> *The U.S. East Coast got pounded by yet another winter storm on Monday that brought temperatures to a 141-year record low in the Baltimore-Washington, D.C. metro area.*[165]

According to *The Weather Channel,*[166] the winter of 2014–15 was "the all-time snowiest" in Boston with "4 of its top 5 snowiest seasons" coming during the previous 21 years.

Record numbers of storms and snowfall totals were recorded in:

- Worcester, Massachusetts (*"February 2015 was their snowiest month, all-time. Incredibly, they've had 2 of their top 5 snowiest months back-to-back,"* January and February.)
- Providence, Rhode Island (*"February 2015 was their second snowiest month, all-time."*)
- Portland, Maine (*"Winter Storm Juno in late January was a top-five snowstorm for Portland."*)
- Bangor, Maine (*"February was the fifth snowiest month and second snowiest February for Bangor."*)
- *"Downeast Maine locations set several 7-to-10 day snowstorm records from late January to early February."*[167]

Further evidence contrary to the prosecution's narrative was the bitter cold European and North African winter of 2016–2017[168] when a rare snowfall, *the first in 40 years,* blanketed the Sahara Desert.[169]

Within the year, a persistent bitter cold end to 2017 and beginning of 2018 set new record cold temperatures in many parts of the US,[170] and the temperature (10°F, -12.2°C) in New York City on New Year's Eve was the *second coldest ever recorded,* making climate alarmist's projections all the more laughable.

Note to politicians: *Be wary of appeals to authority when those "authorities" make a habit of being very, very wrong!*

Not to worry.

The prosecutor's agents were busily rearranging temperature records in early 2018 to "homogenize" the bitter cold right out of the record (Figure B-6).

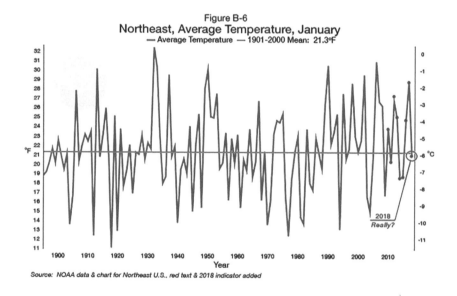

Figure B-6
Northeast, Average Temperature, January
— Average Temperature — 1901-2000 Mean: 21.3°F

Source: NOAA data & chart for Northeast U.S., red text & 2018 indicator added

This was followed in early 2018 by claiming January 2018 had experienced "normal" cold.[71–72, 79, 171]

Homogenization record tampering is covered in Chapter 5.

While it is true such weather events are not climate, they clearly question the validity of the prosecution's theory upon which its climate change models are based. Theory-based models have consistently projected dramatic climate warming that, if true, would make such cold deviations extremely unlikely. This is particularly evident given the extent and frequency of these storms during the winters of 2014–2015 and 2017–2018.

Despite such evidence, recent climate variations are well within normal climate variability. Yet in today's "politically-correct" atmosphere, all challenges to the human-caused climate change dogma are routinely dispatched with a dismissive "the debate is over" or "the science is settled."

Concurrently, the prosecution assails challengers as deniers, sometimes coupled with demands that deniers be imprisoned[172] (or even executed[173]), simply because they have the temerity to question the prosecution's highly dubious climate change theory that itself is in denial of the actual observed evidence in the records!

Based on the testimony of the evidence found in records for carbon dioxide and temperature, *if the debate is over*, it is only because the real-world evidence is unequivocal and so compelling that once this is understood, no rational person could continue to believe the growth of contemporary atmospheric CO_2 concentration has any capacity to discernibly change global climate.

The prosecution's theory of a climate crime is not only unsustained by *the scientific method*, it is invalidated and *must be* rejected.

To be skeptical is the essence of being a good scientist because in the realm of real science, when science is based on theory, it is *never* settled, and it certainly isn't settled by consensus! Only after being subjected to serious skeptical scrutiny, testing, and examination that successfully complies with requirements of *the scientific method* can any theory be considered reliable.

Skepticism, questioning, considering alternatives, and reconsidering theories and hypotheses are all traits fundamental to sound scientific inquiry. Skepticism is both a necessary and healthy part of scientific inquiry because it challenges dogma, inspires reconsideration of paradigms, and promotes development of new theories while advancing science and expanding the sphere of human knowledge.

Perhaps most important, skepticism is the basis for determining compliance with *the scientific method,* and therefore *it is an indispensable part of the scientific process* that scrutinizes theory.

Yet in halls of academia, politics, government agencies, news media, and corporations many are cowed into compliance with the climate change narrative that demands the prosecution's greenhouse gas climate change theory cannot be questioned. It is considered heresy for anyone, particularly a scientist, to have the temerity to even suggest climate change theory be scrutinized by *the scientific method* because, as Mr. Gore and the climate change mob demand, "the science is settled!"

Thoughtful jurors will see through the name-calling, bullying and emotional appeals of a prosecutor who relies on keeping jurors in the dark about key elements of this important global issue.

Jurors should ask, has the prosecution ever:

- Attempted to put contemporary climate change in perspective with Earth's long climate history?
- Admitted to its *lack of understanding* of both past climate and the full spectrum of climate change forces?
- Admitted that an interglacial (current climate) is the *least typical* of any climate regime Earth experiences, and therefore, interglacial climate is not a suitable basis for declaring what climate should be?

- Admitted that it cannot point to any meaningful portion of any untampered CO_2 and temperature records that do not contradict its theorized relationship between the two?

Answer: No…it has not, it cannot, and it will not.

Why? Because its *climate change* case would collapse if it did.

Whenever the phrase "the debate is over" or "the science is settled" is heard, jurors will know they are in the presence of someone who is either ignorant, close-minded, or a charlatan aiming to deceive. Such statements are the antithesis of sound scientific inquiry.

Summary: Additional Topics Supporting Defense Testimony

1. Human activity's CO_2 emissions do not impact ocean pH. Ocean acidification is not a legitimate concern.
2. Based on the clear evidence from records of atmospheric CO_2 and global average surface temperature, changing atmospheric CO_2 is unrelated to climate change.
3. Cow flatulence (methane) has no impact on climate change, regardless of the prosecution's silly claims to the contrary.
4. Solar variability, orbital cycles and geologic activity are known natural climate change forces capable of creating climate variability.
5. All sources of historic climate variability are not known with certainty.
6. The testimony of actual real-world evidence shows perfectly normal climate variability characteristic of routine climate change since the end of the Little Ice Age, with recent warming having been a likely manifestation of the solar grand maximum that peaked in the late 20th-century.

7. The prosecution routinely supports its case with a host of deceptions designed to mislead the jury.
8. While prosecution agents continue to claim human-caused climate change poses an "existential threat" to the future of humanity, there is zero evidence to support that wild speculation.
9. The prosecution's case is entirely built on theory, not evidence. As *the scientific method* demands, when theory is contradicted by real-world evidence, believe the real world evidence and reject the invalid theory.

DEFENSE EXHIBIT C

Summary of Key Findings in Chapters

Derived from the materials presented in Chapters 1 through 5, the following material is a partial list of the key evidence brought out in testimony for the defense:

- To the extent climate is changing, the observed change is well within the bounds of perfectly natural climate change. Contemporary climate change is neither unprecedented nor unusual.

- Examination of official US government records for global averages of both atmospheric CO_2 and surface temperature (GAST) during the 139 years (1880 through 2018) of recorded measurements shows no theory-consistent response of either temperature or climate (30 years of temperature change) to changes in global atmospheric CO_2 (see examples in Defense Exhibit D).

- There is no correlation between annual changes in atmospheric CO_2 and the annual change of global average temperatures (land, ocean, or merged ocean plus land) over (1) the 139-year period from 1880 through 2018, (2) the 130-year period from 1885 through 2014, (3) the 75-year period from 1940 through 2014, (4) the 60-year period from 1951 through 2010, or (5) the most recent fifteen years (2005-2019) during which USCRN land temperature monitoring began. Without correlation there can be no causation.

- Two non-overlapping 41-year profiles of 20th-century climate change, 1904 to 1944 and 1958 to 1998, cannot be differentiated based on their visual appearance. While climate slightly warmed in both periods, the most recent period actually warmed *less than* the earlier period by 0.19°C (0.34°F) while atmospheric CO_2 was growing during the latter period by 404% *more than* during the earlier period. Climate change during the 20th-century was unremarkable and representative of a perfectly natural overall slow warming trend since the end of the Little Ice Age in the mid-19th-century. Prosecution claims to the contrary are simply not supported by the records.

- Global average atmospheric CO_2 and global average temperature are strongly uncorrelated (r = 0.29) over 25 million year increments of the past 550 million years of geologic evidence. Global average atmospheric CO_2 change and global average temperature change are very strongly uncorrelated (r = 0.10) over 25 million year increments of the past 550 million years. Causation cannot exist without strong correlation. If atmospheric CO_2 change (growth and decline) were a meaningful climate change force, it would necessarily be strongly correlated with global climate change over every meaningful timeframe, but the two are strongly uncorrelated (r = 0.10) over the past 550 million years.

- Earth's climate and atmospheric CO_2 histories testify to there being no evidence of any relationship that supports the claim that atmospheric CO_2 growth causes climate to warm.

- Whether warming or cooling, over the past 550 million years, no measured significant climate change event can be shown to have been a consequence of either a corresponding atmospheric CO_2 growth or a corresponding CO_2 decline.

- Climate change has been shown to be insensitive to total atmospheric CO_2 change over every time frame examined. Therefore, all sources of anthropogenic CO_2 emissions, particularly those attributed to fossil fuel use, are entirely irrelevant and have no impact whatsoever on Earth's climate.

- The United States, as the only major G20 nation having withdrawn from the *Paris Agreement*, is the only nation basing its policy on well-founded prudent economic and environmental considerations dictated by credible scientific evidence. Contrary beliefs are not supported by real-world evidence and rest on a shaky foundation of misinformation, unscientific speculative assertions, questionable conduct, a dubious theory, and actions designed to deceive the public.

- The *Paris Agreement* is a costly fiasco *that can have no impact on climate* and, therefore, serves neither a legitimate environmental purpose nor any other constructive purpose. The United States was foolish to hastily enter into the ill-conceived agreement and wise to correct that foolishness by withdrawing. Other nations would be wise to follow.

- No evidence of significant climate change exists to justify costly premature development of non-fossil fuel energy schemes. Prudence, not pretense or fear, should dictate a measured development of intelligent, affordable alternate energy sources to the finite supply of fossil fuel.

- An awareness of inevitable future diminishing supplies of fossil fuels suggests a sensible course would include prudent privately-funded competitive research and development of economically feasible alternate energy sources, preferably renewable but certainly capable of producing sufficient affordable energy to meet the realistic needs of a dynamic civilization embracing a free-market economy. Such efforts should not be approached in haste motivated by unjustifiable concerns about nonexistent human-induced climate change.

- Those scientists who, in defiance of the compelling evidence to the contrary, still cling to the belief that atmospheric CO_2 is a major climate change force, would be well-advised to carefully examine the well-established record of atmospheric CO_2 change and global average temperature change and then explain how two completely uncorrelated factors can possibly have the highly correlated relationship demanded by the greenhouse gas climate change theory.

- Climate scientists would be well-advised to concentrate their research on a better understanding of *natural climate change* and the natural forces that control significant cold excursions (e.g., ice eras, ice epochs, and ice age-interglacial cycles) from Earth's typically much warmer climate (far warmer than any climate humans have ever experienced).

- Persistent references to fossil fuel burning being responsible for climate change are entirely unfounded and constitute a completely false narrative that deceives the public.

- Branding carbon dioxide a pollutant is dishonest, inappropriate, misguided, and unfounded; carbon dioxide is an essential gas for all carbon-based life on Earth. Below 135 ppm, atmospheric CO_2 is too low to sustain carbon-based life. Current atmospheric CO_2 is historically low and well below levels that existed when plant life evolved (at least 800 ppm for most plants). From the perspective of plant life, Earth's atmosphere is CO_2-starved.

- Costly advances in alternate energy technology should be borne by private entrepreneurs using unsubsidized private capital to competitively develop the technology required to make alternate energy concepts economically cost-effective.

- Humanity is best served by preparing for both inevitable global climate warming and cooling, both of which are certain in Earth's future and *cannot be averted*.

- Future *global cooling* poses a greater existential threat than does global warming. As the Holocene interglacial ends, ice age cooling will create a much greater threat to humanity than global warming. Less than fifteen thousand years ago near the end of the last ice age glacial advance, a mile-thick continental glacier covered all of Canada, all of New England, New York State, northern New Jersey, Pennsylvania, and much of Ohio where the glacier angled southwest to a latitude below that of Denver, Colorado, and covering virtually all of the upper midwest before angling northwestward to below Bismarck, North Dakota, and then in a generally westward course to the northern coast of Washington. There is no reason to believe future ice age glacial ice will not extend at least as far south as the last. In terms of significant climate change, Earth is on the doorstep of the next ice age cycle.

- At recent G20 Summits, without any serious consideration of the adverse consequences of doing so, member nations routinely reaffirm their commitment to abide by the base-less fossil fuel emissions restrictions of the *Paris Agreement*. That agreement is motivated by the false premises that (1) the dominant contributor to atmospheric CO_2 growth is humanity's use of fossil fuels (coal, oil, natural gas) and (2) atmospheric CO_2 growth is causing climate to warm, both of which are specious nonsense that lack any supporting evidence in observed records. The *Paris Agreement* has been shown to be spectacularly misguided and unsupported by the clear record of changing atmospheric CO_2 and global average climate change over every time frame examined from contemporary records, and ice core analyses, to geologic-based climate and atmospheric CO_2 reconstruction over the past half-billion years.

- Testimony presented to the jury has unequivocally demonstrated that observed temperature and climate change is insensitive to observed atmospheric CO_2 change over (1) the contemporary years between 1880 and 2018; (2) the GISP2 ice core analysis of the 10,700 years of the Holocene interglacial; (3) analyses of the contradictory temporal relationship to theory (climate changes first, *followed* by CO_2 change) over 800,000 years of Antarctic ice core evidence; and (4) the most recent 550 million years of the geologic record.

- At every opportunity purveyors of the unsupportable climate change narrative irresponsibly demonize fossil fuels, claiming failure to restrict their emissions is likely to create potentially catastrophic human-induced climate change. Considering only a small portion of atmospheric CO_2 arises from fossil fuel use, there is no credible evidence to support that narrative; to the contrary, the best available observed evidence strongly contradicts the prosecution's climate change narrative.

- By following credible evidence, not invalid theory, the jury is compelled to declare both anthropogenic carbon dioxide and total atmospheric carbon dioxide innocent of all charges relating to climate change. Using US government records for global atmospheric CO_2 and global temperature, the defense has unequivocally produced compelling evidence that all claims of human-induced global climate change are without any scientific basis because they are dramatically contradicted across the spectrum of evidence scrutinized.

- Climate change concerns about carbon footprints, carbon pollution, and fossil fuel energy are folly based on unscientific speculative misuse of theory ultimately found to be invalid. Such concerns can be safely ignored because neither human activity's emissions of CO_2 nor total atmospheric CO_2 change from any other source have a discernible impact on global climate change. Fears of catastrophic human-induced climate change are baseless and rooted in an invalid and deeply-flawed theory that the jury has seen to be pure scientific fantasy.

- Using *the scientific method*, the defense has offered the jury compelling evidence that the greenhouse gas climate change theory is fundamentally flawed and invalid for contradicting observed evidence of the relationship between global atmospheric CO_2 change and global average temperature change. Failing *any* test of *the scientific method* is a sufficient basis for rejecting the prosecution's greenhouse gas climate change theory as invalid, rendering it little more than unsupportable speculative nonsense.

- Nations must focus on real problems and ignore phantom fears manufactured around a deceptive specious theory that demonizes humanity's most affordable, reliable, and suitable source of energy to power modern civilization.

- Clean energy initiatives should be intelligently developed with private funding. Those who seek to develop new energy sources should be willing to risk their own capital; they should not rely on a false climate change narrative

that stampedes politicians to fund grants from the public treasury to billionaire-backed syndicates seeking to profit from public hysteria over a complete fabrication the likes of which has not been seen since the Piltdown Man in 1912 ("Piltdown Man was an audacious fake and a sophisticated scientific fraud."[121]).

- The failure of contemporary journalism to expose weaknesses in the prosecution's case and the strengths of the defense case suggest a greater desire for profit from fearmongering than a commitment to honestly informing the public, a growing characteristic of modern journalism-for-profit (a.k.a. "fake news"). Television and newspaper stories pandering to the prosecution have grossly misled the public and for that such reports are nothing short of dishonest fearmongering.

- The failures of once-respected scientific professional organizations to pursue an unbiased quest for truth have tarnished these organizations as never before, not only in the eyes of the public but also in the view of their membership.

- The meek acceptance and furtherance of the false climate change narrative by public schools and institutions of higher learning threaten irreparable damage to science and the body of degreed graduates who have been misled to believe that the prosecution's arguments are scientifically valid when they are not.

- No *credible evidence* exists to support any claim that global climate change is caused either by atmospheric CO_2 change or by any human activity.

- Compelling evidence that the prosecution's theory is invalid is summarized in Defense Exhibits D-3, D-4, and D-5.

DEFENSE EXHIBIT D

Six Theory-Challenging Exhibits

DEFENSE EXHIBIT D-1

A Climate Change Quiz

Four multiple-choice questions. How well will you do?

1. How many of the 50 US states recorded their *all-time record high* temperature during or after 1960?

    (a.) 41

    (b.) 35

    (c.) 22

    (d.) 14

2. How many of the 50 US states recorded their *all-time record low* temperature during or after 1960?

    (a.) 31

    (b.) 22

    (c.) 14

    (d.) 7

3. How many of the 50 US states recorded their *all-time record high* temperature during or after 2000?

    (a.) 12

    (b.) 8

    (c.) 5

    (d.) 2

4. How many of the 50 US states recorded their *all-time record low* temperature during or after 2000?

    (a.) 9

    (b.) 6

    (c.) 3

    (d.) 0

Answers on next page.

Answers: 1. (d) 2. (b) 3. (d) 4. (c)

The More Things Change, the More They Stay the Same

The hot dry seasons of the past few years have caused rapid disintegration of glaciers in Glacier National Park, Montana… Sperry Glacier…has lost one-quarter or perhaps one-third of its ice in the past 18 years… If this rapid rate should continue…the glacier would almost disappear in another 25 years…

Born about 4,000 years ago, the glaciers that are the chief attraction in Glacier National Park are shrinking so rapidly that a person who visited them ten or fifteen years ago would hardly recognize them today as the same ice masses.

Do these reports sound familiar? Typical of frequent warnings of the dire consequences to be expected from global warming, such reports often refer to modern civilization's use of fossil fuels as being the dominant cause of recent climate change.

You might be surprised to learn the reports above were made nearly thirty years apart! The first in 1923 *prior to* the record heat of the Dust Bowl years during the 1930s. The second in 1952 during the second decade of a four-decade *cooling trend* that had some scientists concerned that a new ice age might be on the horizon!

Did the remnants of Sperry Glacier disappear during global warming of the late 20th century?

According to the US Geological Survey (USGS), today Sperry Glacier "ranks as a moderately sized glacier" in Glacier National Park.

What caused the warmer global climate prior to "4,000 years ago" before Glacier National Park's glaciers first appeared?

Are you aware that during 2019 the National Park Service quietly began removing its "Gone by 2020" signs from Glacier National Park as its most famous glaciers continued their renewed growth that began in 2010?

Was late 20th-century global warming caused by fossil fuel emissions? Was it really more pronounced than early 20th-century warming? Or was late 20th-century warming perfectly natural, in part a response to the concurrent peak strength of one of the strongest solar grand maxima in a very long time?

2005–2019 NOAA, USCRN Evidence

To obtain an unbiased distribution of records across various terrains and climate zones, the United States Climate Reference Network (USCRN, NOAA) established a network of 114 new carefully-sited and maintained temperature and precipitation monitoring stations distributed throughout the 48 contiguous states. Stations were sited so that no unnatural environmental influences could taint the records. When completed, this network will include both Hawaii and Alaska.

Operational in 2005, these USCRN records provide the best evidence for early 21st-century land surface temperature in the contiguous 48 states. Daily records are used to create monthly histories (January to December). For each year, the US monthly records are averaged to produce a continent-wide yearly temperature record since 2005. According to the USCRN network, during the 14 years between 2005 and 2019, while atmospheric CO_2 *grew by 30 ppm*, contrary to theory, *a net cooling of -0.41°C (-0.74°F)* was observed.

When a 30 ppm *growth* of atmospheric CO_2 occurs during the same 14 years when temperature *cooled*, any theory that claims growing atmospheric CO_2 will cause temperatures to warm is contradicted by the recorded evidence in nature.

USCRN evidence recently began to see the light of public scrutiny. From a July 30, 2019 article at *Watts Up With That?* by meteorologist Anthony Watts:

> While media outlets scream "hottest ever" for the world in June and July (it's summer) and opportunistic climate crusaders use those headlines to push the idea of a "climate crisis" the reality…for [the] USA is that so far most of 2019 [temperatures have] been below normal…

> Little known data from the state of the art [USCRN] (which never seems to make it into NOAA's monthly "state of the climate" reports) show that for the past nine months, six of them were below normal…

All while yearly atmospheric CO_2 has been relentlessly growing since 1944. Between 2005 and 2019, seven months cooled while five months warmed, creating the net 14-year cooling of -0.41°C (-0.74°F). It should be noted that the discontinuity between atmospheric CO_2 change and climate is dramatic in the fifteen years of the USCRN records. While annual CO_2 growth averaged 2.26 ppm per year, the average annual temperature cooled by -0.03°C (-0.05°F) per year, just the opposite of what climate change theory demands.

Questions to ponder: Why is this evidence that contradicts climate change theory being ignored by major news outlets? When the best evidence available across the contiguous 48 states defies climate change theory, which should be believed: (1) climate change *theory* or (2) real-world *evidence*?

DEFENSE EXHIBIT D-4

1912–1976 NOAA, NCDC, MLO Evidence

Greenhouse gas (GHG) climate change theory asserts that as atmospheric CO_2 changes over time, temperatures will change accordingly. If atmospheric CO_2 grows, climate will warm; if CO_2 declines, climate will cool.

Does the observed evidence in nature support this theorized relationship?

NOAA maintains yearly records for global average atmospheric CO_2 and merged land plus ocean global average surface temperatures (GAST) going back to 1880. These records are the observed evidence for what has occurred in nature.

It is generally acknowledged that thirty years of observed temperature change is sufficient to constitute evidence of climate change.

During the 32 years from 1912 through 1944, NOAA records show atmospheric CO_2 modestly grew by 9.2 ppm while 32 years of GAST (climate) showed a pronounced warming of 0.55°C (nearly 1°F).

Yet during the 32 years from 1944 through 1976, NOAA's records show atmospheric CO_2 grew by 21.84 ppm or *more than twice* its growth during the earlier 32 years. Yet, while atmospheric CO_2 was skyrocketing, climate *cooled* by -0.37°C (-0.66°F).

To summarize, for 32 years *Earth's climate sharply warmed during very modest atmospheric CO_2 growth.* For 32 years during which *Earth's climate persistently cooled, atmospheric CO_2 grew dramatically more (237% greater)* than during the 32 years of *sharp warming!* Contrary to GHG climate change theory, the strong growth of atmospheric CO_2 accompanied *cooling* while slight CO_2 growth accompanied *dramatic warming!*

These NOAA records reveal the theorized strong relationship between climate change and atmospheric CO_2 change is contradicted by the observed evidence in nature. This constitutes compelling evidence that something is very wrong with the greenhouse gas climate change theory.

DEFENSE EXHIBIT D-5

1880–2018 NOAA, NCDC, MLO Evidence

Greenhouse gas (GHG) climate change theory asserts that as atmospheric CO_2 changes over time, temperatures will change accordingly. If CO_2 increases, temperatures warm; if CO_2 decreases, temperatures cool.

Does the observed evidence in nature agree?

NOAA maintains yearly records for global average atmospheric CO_2 and merged land plus ocean global average surface temperature (GAST) going back to 1880. These records are the observed evidence for what has occurred in nature.

Using the January 2019 NOAA records spanning 139 years of atmospheric CO_2 and GAST from 1880 through 2018, the observed evidence can be examined to see whether the greenhouse gas climate change theory is supported by the evidence in nature.

During the 65 years from 1880 through 1944, atmospheric CO_2 grew by *19 ppm* (from 291 ppm to 310 ppm), an average growth of 0.29 ppm/year. During the 74 years from 1945 through 2018, atmospheric CO_2 grew by *98 ppm* (from 310 ppm to 408 ppm), an average increase of 1.32 ppm/year, or *more than four-and-one-half times* the yearly CO_2 growth rate during the prior 65 years from 1880 through 1944.

In accordance with climate change theory, did Earth's surface temperatures respond to the dramatic increase in atmospheric CO_2 after 1944?

During the 65 years from 1880 through 1944, GAST warmed by 0.42°C, an average warming of 0.0065°C/year. During the 74 years of dramatic atmospheric CO_2 growth from 1945 through 2018, GAST warmed by 0.50°C, an average warming of 0.0068°C/year, *virtually identical* to the average for the prior 65 years (to three decimal places, they *are* identical).

Contradicting climate change theory, the records show that despite a dramatic increase in the rate of atmospheric CO_2 growth after 1944, the impact on observed temperature change was inconsequential.

Conclusion: The 139 years (1880–2018) of NOAA's records for atmospheric CO_2 and GAST provide compelling evidence refuting the greenhouse gas climate change theory's asserted relationship between atmospheric CO_2 growth and GAST.

DEFENSE EXHIBIT D-6

A Sweet Analogy

This visualization relates grains of table sugar to molecules of Earth's atmospheric gases (nitrogen, oxygen, etc.):

One level tablespoon of granulated (table) sugar contains about 54,408 granules of sugar. Imagine these granules represent every molecule of Earth's atmospheric gases. Just 22 of the sugar granules, randomly spread throughout the sugar, represent Earth's *total* atmospheric CO_2. Twenty of those granules are tinted a pale green while two of them are tinted orange.

The pale green granules represent *total* atmospheric carbon dioxide from *natural sources* (not related to any human activity). The two orange granules represent the prosecution's exaggerated estimate of the contribution to atmospheric CO_2 from accumulated fossil fuel combustion during the entire 20th century. Note that according to many other scientists, the actual fossil fuel contribution is realistically about *one-tenth* the amount alleged by the prosecution (less than a quarter of one granule!). For the purpose of this visualization, the defense stipulates to the much higher figure preferred by the prosecution.

Suppose the aggregate temperature of these 54,408 sugar granules is 59°F. The prosecution's climate change theory alleges the addition of *just those two orange sugar granules*, and the prospect of a few additional granules accumulating is sufficient to "catastrophically" raise the aggregate temperature of the other 54,406 sugar granules. Do any jurors seriously believe that those two granules (or the more realistic

one-quarter of one granule) can make any noticeable change to either the sugar's temperature or its perceived color? Does anyone seriously believe the sugar's temperature would noticeably change if those two orange granules were removed? Or doubled to four? Would the sugar's temperature noticeably change if the temperature of those two orange granules were 65°F, or 70°F, or even 80°F?

Those two granules out of 54,408 form the foundation for today's climate change hysteria. Without those two orange granules, there would be no dubious climate change theory to invalidate.

REFERENCES AND DATA SOURCES

Please note: Links to references often change over time. If the published link no longer works, search on the title to find the material. Add author(s) to search to help refine results.

1. ESS-DIVE, (2014), *Global CO$_2$ Emissions from Fossil-Fuel Burning, Cement Manufacture, and Gas Flaring: 1751-2014*, (archives for Carbon Dioxide Information Analysis Center, CDIAC), Annual anthropogenic CO$_2$ emissions, (https://data.ess-dive.lbl.gov/view/doi:10.3334/CDIAC/00001_V2017).

2. IPCC, (2014), *Climate Change 2014 Synthesis Report, Summary for Policymakers*, page 4, SPM 1.2, (https://www.ipcc.ch/pdf/assessment-report/ar5/syr/AR5_SYR_FINAL_SPM.pdf).

3. NOAA Earth System Research Laboratory (ESRL), Global Monitoring Division, *Trends in Atmospheric Carbon Dioxide*, monthly CO$_2$ records, (https://www.esrl.noaa.gov/gmd/ccgg/trends/monthly.html).

4. NOAA Earth System Research Laboratory (ESRL), *Mauna Loa global CO$_2$ growth* (ftp://aftp.cmdl.noaa.gov/products/trends/CO2/CO2_annmean_mlo.txt, and NOAA Earth System Research Laboratory at https://www.esrl.noaa.gov/gmd/ccgg/trends/data.html).

5. NOAA, (July 13, 2018), *NOAA Monthly CO$_2$ Data*, (ftp://aftp.cmdl.noaa.gov/products/trends/CO2/CO2_mm_gl.txt).

6. NOAA Earth System Research Laboratory (ESRL), Global Monitoring Division, *Trends in Atmospheric Carbon Dioxide*, annual global, (https://www.esrl.noaa.gov/gmd/ccgg/trends/global.html).

7. NCDC, (2017), *ERSST-v4 monthly and yearly temperature anomaly estimates, 1880–2017*, (ftp://ftp.ncdc.noaa.gov/pub/data/noaaglobaltemp/operational/timeseries/).

8. Australian Government, *Energy Smart* (pamphlet), (http://www.whitehorse.vic.gov.au/IgnitionSuite/uploads/docs/Sustainable%20Living%20Guide%20Energy.pdf).

9. Schneider, Steven, (1978), *In Search of the Coming Ice Age*, YouTube video, (https://www.youtube.com/watch?v=aA-6JbGbs3s).

10. The New England Historical Society, (updated, 2018), *1816: The Year Without a Summer*, (https://www.newenglandhistoricalsociety.com/1816-year-without-a-summer/).

11. Wikipedia, *Little Ice Age*, (https://en.wikipedia.org/wiki/Little_Ice_Age).

12. IPCC, (October 2013), *Principles Governing IPCC Work*, (http://www.ipcc.ch/pdf/ipcc-principles/ipcc-principles.pdf).

13. D'Aleo, Joseph, Dwyer, Jr., Clement, Slanover, Russell C., Univer, Scott M., Wallace III, James P., Weaver, Robin D. and Springer, Douglas S., (2018), *Fifth Supplement To Petition For Reconsideration Of "Endangerment And Cause Or Contribute Findings For Greenhouse Gases Under Section 202(a) Of The Clean Air Act,"* (https://thsresearch.files.wordpress.com/2018/02/ef-cpp-fifth-supplement-to-petition-for-recon0d0a-final020918.pdf).

14. D Kriebel, et al, (September 2001), *The precautionary principle in environmental science,* (https://www.ncbi.nlm.nih.gov/pmc/articles/PMC1240435/).

15. Wikipedia, *Daniel Webster,* https://en.wikipedia.org/wiki/Daniel_Webster).

16. USA Today Editorial Board, (July 30, 2018), *"A hellish July validates climate change forecasts,"* (https://www.usatoday.com/story/opinion/2018/07/30/climate-change-hellish-july-validates-forecasts-editorials-debates/855435002/).

17. USGS, (2019), *Sperry Glacier,* (https://www2.usgs.gov/climate_landuse/clu_rd/glacierstudies/sperry.asp).

18. Naval History Blog (US Naval Institute), (August 11, 2011), *USS Skate (SSN-578) Becomes the First Submarine to Surface at the North Pole,* (https://www.navalhistory.org/2011/08/11/uss-skate-ssn-578-becomes-the-first-submarine-to-surface-at-the-north-pole).

19. Shabecoff, Philip, (January 29, 1989), *U.S. Data Since 1895 Fail to Show Warming Trend,* The New York Times, (https://www.nytimes.com/1989/01/26/us/us-data-since-1895-fail-to-show-warming-trend.html).

20. Editorial, (March 29, 2018), *The Stunning Statistical Fraud Behind The Global Warming Scare,* Investor's Business Daily, (https://www.investors.com/politics/editorials/the-stunning-statistical-fraud-behind-the-global-warming-scare/).

21. Marohasy, Jennifer, (August 21, 2017), *Most of the Recent Warming Could be Natural,* Jennifer Marohasy website, (http://jennifermarohasy.com/2017/08/recent-warming-natural/).

22. Lindzen, Richard, (April 17, 2018), *Classic Lecture on Climate Sensitivity,* YouTube, (https://www.youtube.com/watch?v=H2czGg3fUUA).

23. U. of Indiana, *Milankovitch Cycles and Glaciation,* (https://www.indiana.edu/~geol105/images/gaia_chapter_4/milankovitch.htm) and "Milankovitch cycles," Wikipedia, (https://en.wikipedia.org/wiki/Milankovitch_cycles).

24. Kamis, James Edward, (May 14, 2015), *Emerging 2015 El Nino Fueled By Deep Ocean Geological Forces,* Plate Climatology, (http://www.plateclimatology.com/emerging-2015-el-nino-fueled-by-deep-ocean-geological-forces/).

25. Richard, Kenneth, (April 11, 2017), *Modern Solar Grand Maximum Ends: 'Little Ice Age' Cooling Coming!,* Principia Scientific International, (https://principia-scientific.org/modern-solar-grand-maximum-ends-little-ice-age-cooling-coming/).

26. "Planet Earth," (1983), pages 162–169, *Continents in Collision,* book series by Time-Life Books, Inc.

27. Dvorsky, George, (May 8, 2018), *Are Jupiter and Venus Messing With Earth's Climate?,* Gizmodo, (https://gizmodo.com/are-jupiter-and-venus-messing-with-earth-s-climate-1825858316).

28. Wikipedia, *Isotope,* (https://en.wikipedia.org/wiki/Isotope).

29. Wikipedia, *Water Vapor,* graph associated with "Impact on air density," (https://en.wikipedia.org/wiki/Water_vapor).

30. Universe Today, Carbon Atom illustration, (https://www.universetoday.com/56469/atom-diagram/).

31. Segalstad, Dr. Tom V., (1998), *Carbon cycle modelling and the residence time of natural and anthropogenic atmospheric CO_2: on the construction of the 'Greenhouse Effect Global Warming' dogma,* In Bate, R. (Ed.): Global warming: the continuing debate. ESEF, Cambridge, U.K. (ISBN 0952773422), pp. 184–219, (http://www.CO2web.info/ESEF3VO2.pdf).

32. Hermann Harde, (May 2017), *Scrutinizing the carbon cycle and CO_2 residence time in the atmosphere,* Journal: Global and Planetary Change, (http://www.sciencedirect.com/science/article/pii/S0921818116304787).

33. Glassman, Jeffrey A. Ph D, 2010, *CO_2: "Why Me?"—On Why CO_2 is Known NOT to Have Accumulated in the Atmosphere & What is Happening with CO_2 in the Modern Era,* Rocket Scientist's Journal, (http://www.rocketscientistsjournal.com/2007/06/on_why_CO2_is_known_not_to_hav.html).

34. Segalstad, T.V., (August 8, 2008), *Carbon isotope mass balance modelling of atmospheric vs. oceanic CO_2,* 33rd International Geological Congress, Norway, Program & Abstracts (PDF; 57 kBytes), (http://www.CO2web.info/Segalstad_C-isotopes_33IGC.pdf).

35. NTNU, (December 4, 2013), *Unlocking the secrets of marine carbon cycling,* Norwegian University of Science and Technology (NTNU), (https://www.ntnu.edu/news/2013-news/ocean-certain).

36. Segalstad, Tom, D., (January 2014), *Some Thoughts on Ocean Chemistry,* Climate Change Reconsidered II—Biological Impacts, Chapter: 6.3.1.2, (https://www.researchgate.net/profile/Tom_Segalstad/publication/304797201_Some_thoughts_on_ocean_chemistry_Chapter_6312/links/577b9f7608ae355e74f15a80/Some-thoughts-on-ocean-chemistry-Chapter-6312.pdf?origin=publication_detail).

37. Dockery, Bevan, (January 19, 2018), *Shock Study: Atmospheric CO_2 Levels Change With Planetary Movements,* Principia Scientific International, (https://principia-scientific.org/shock-study-atmospheric-CO2-levels-change-with-planetary-movements/).

38. "Planet Earth," (1983), pages 20–21, *Ice Ages,* book series by Time-Life Books, Inc.

39. (2001), *Climate and the Carboniferous Period,* Temperature after C.R. Scotese (http://www.scotese.com/climate.htm), CO_2 after R.A. Berner, 2001 (GEOCARB III), (http://www.geocraft.com/WVFossils/Carboniferous_climate.html).

40. Reviews of Geophysics (An American Geophysical Union Journal), (March 5, 2016), *Interglacials of the last 800,000 years,* (http://onlinelibrary.wiley.com/doi/10.1002/2015RG000482/full).

41. Hooker, Dolph Earl, (1958), *Those Astounding Ice Ages—An exploration of our planet's most challenging mysteries,* by Exposition Press, New York.

42. "Planet Earth," (1983), pages 157–159, *Atmosphere,* book series by Time-Life Books, Inc.

43. NASA Earth Observatory, *Milutin Milankovitch (1879-1958),* (https://earthobservatory.nasa.gov/features/Milankovitch/milankovitch_2.php).

44. Wikipedia, *Orbital Eccentricity,* (https://en.wikipedia.org/wiki/Orbital_eccentricity).

45. IPCC, (2014), *Climate Change 2014 Synthesis Report, Summary for Policymakers,* page 5, SPM 1.2, (https://www.ipcc.ch/site/assets/uploads/2018/02/AR5_SYR_FINAL_SPM.pdf).

46. Soon, William, (2015), *Climate Change: The Facts/Chapter 4—Sun shunned,* Institute of Public Affairs & Stockade Books (edited by Alan Moran).

47. Soon, William, (2015), *Climate Change: The Facts/Chapter 4—Sun shunned,* (footnotes 17–21, pages 293–295), Institute of Public Affairs & Stockade Books (edited by Alan Moran).

48. Soon, W. and Legates, D.R., (2013), *Solar irradiance modulation of equator-to-pole (Arctic) temperature gradients: Empirical evidence for climate variation on multidecadal timescales,* Journal of Atmospheric and Solar-Terrestrial Physics, Vol. 93 (2013), 45–56.

49. Stott, Peter A., Jones, Gareth S., Mitchell, John F. B., (June 10, 2003), *Do Models Underestimate the Solar Contribution to Recent Climate Change?*, Hadley Centre for Climate Prediction and Research, Met Office, Bracknell, Berkshire, UK, (http://climate.envsci.rutgers.edu/pdf/StottEtAl.pdf).

50. Hertzberg, Martin and Schreuder, Hans, 2017, *Role of greenhouse gases in climate change,* (http://tech-know-group.com/papers/Role of CO2-EaE.pdf).

51. Kamis, James Edward, (February 24, 2015), *Rift System Geology Rocks Global Warming Theory*, Plate Climatology, http://www.plateclimatology.com/rift-system-geology-rocks-global-warming-theory/).

52. Kamis, James Edward, (December 31, 2015), *What's Fueling the Current El Niño (and It's Not Global Warming)*, Plate Climatology, (http://www.plateclimatology.com/whats-fueling-the-current-el-nino-and-its-not-global-warming/).

53. Rice, K., (June 18, 2017), *What does the Vostok ice core tell us?*, (https://andthentheresphysics.wordpress.com/2017/06/18/what-does-the-vostok-ice-core-tell-us/).

54. Latour, Dr Pierre, (June 13, 2014), *Atmospheric Carbon Dioxide Lags Temperature: The Proof*, Principia Scientific International, (https://principia-scientific.org/atmospheric-carbon-dioxide-lags-temperature-the-proof/).

55. Brehmer, Carl, (3 October 2017), *SURFRAD Data Falsifies the 'Greenhouse Effect' Hypothesis,* (http://tech-know-group.com/essays/SURFRAD Data Falsifies GHE.pdf).

56. Hertzberg, Dr. Martin; Siddons, Alan and Schreuder, Hans, (2017), *Role of greenhouse gases in climate change,* (http://tech-know-group.com/papers/Role of GHE-EaE.pdf).

57. Gerlich, Gerhard, (2007), *Falsification of the Atmospheric CO_2 Greenhouse Effects within the Frame of Physics,* Institute of Mathematical Physics, Technical University Carolo-Wilhelmina (http://citeseerx.ist.psu.edu/viewdoc/download?doi=10.1.1.574.5131&rep=rep1&type=pdf).

58. Miatello, Alberto, (2012), *Refutation of the "Greenhouse Effect" Theory on a Thermodynamic and Hydrostatic Basis,* Principia Scientific International, (https://principia-scientific.org/images/stories/pdfs/PSI Miatello Refutation GHE.pdf).

59. Thieme, Heinz, (2002), *Greenhouse Gas Hypothesis Violates Fundamentals of Physics,* (http://ilovemycarbondioxide.com/archives/Greenhouse Gas Hypothesis Violates Fundamentals of Physics2.pdf).

60. Thieme, Heinz, (2002), *Does Man really affect Weather and Climate?,* (http://realplanet.eu/Influence.htm).

61. Thieme, Heinz, (2002), *On the Phenomenon of Atmospheric Backradiation,* (http://realplanet.eu/backrad.htm).

62. Gerlich, Gerhard, (January 6, 2009), *Falsification Of The Atmospheric CO_2 Greenhouse Effects Within The Frame Of Physics,* (https://arxiv.org/pdf/0707.1161.pdf).

63. Leroux, Professor Marcel PhD, Climatology, (2005), *Global Warming—Myth or Reality? The Erring Ways of Climatology,* Springer-Praxis Books in Environmental Science.

64. Nahle, Prof. Nasif, (2011), *Observations on "Backradiation" during Nighttime and Daytime*, Principia Scientific International, (https://principia-scientific.org/publications/New Concise Experiment on Backradiation.pdf).

65. NOAA, 2014, *Global Climate Report—Annual 2014,* (https://www.ncdc.noaa.gov/sotc/global/201413).

66. NOAA, NCDC, (2019), *Global Average Surface Temperature Anomalies (Annual), 1880-2019,* (https://www.ncdc.noaa.gov/data-access/marineocean-data/noaa-global-surface-temperature-noaaglobaltemp).

67. USA Today, (February 27, 2019, updated), *"99.9999 percent chance we're the cause of global warming, study says,"* based on *Nature* article by Benjamin Santer, et al, (https://www.usatoday.com/story/news/nation/2019/02/26/global-warming-99-9999-percent-chance-humans-cause/2994043002/).

68. climate4you.com, (September 2018), *An overview to get things into perspective*, (http://www.climate4you.com/GlobalTemperatures.htm).

69. Nova, Joanne, (April 2018), *Climate change means Greenland is the same temperature now as 1880*, see also comment #4, JoNova website, (http://joannenova.com.au/2018/04/greenland-same-temperature-now-as-1880/).

70. Time Magazine, (June 1974), *Another Ice Age?*, (article), (https://seeker401.wordpress.com/2009/12/10/time-magazine-june-1974-another-ice-age/).

71. Spencer, Dr. Roy, (October 2017), *UAH Satellite-Based Temperature of the Global Lower Atmosphere* (Version 6.0), (http://www.drroy-spencer.com/wp-content/uploads/UAH_LT_1979_thru_October_2017_v6.jpg).

72. Spencer, Dr. Roy, *Latest Global Average Tropospheric Temperatures*, (http://www.drroyspencer.com/latest-global-temperatures/).

73. Lewis, Mario, (February 6, 2016), *Satellites and Global Warming: Dr. Christy Sets the Record Straight*, (http://www.globalwarming.org/2016/02/05/satellites-and-global-warming-dr-christy-sets-the-record-straight/).

74. Spencer, Roy W.; Christy, John R. and Braswell, William D., (November 2016), *UAH Version 6 Global Satellite Temperature Products: Methodology and Results*, (http://www.drroyspencer.com/wp-content/uploads/APJAS-2016-UAH-Version-6-Global-Satellite-Temperature-Products-for-blog-post.pdf).

75. Radiosonde Museum of North America, *What is a Radiosonde?* (http://radiosondemuseum.org/what-is-a-radiosonde/).

76. Luers, James K., Eskridge, Robert E., (May 1, 1998), *Use of Radiosonde Temperature Data in Climate Studies*, Journal of the American Meteorological Society, (https://journals.ametsoc.org/doi/full/10.1175/1520-0442(1998)011%3C1002%3AUORTDI%3E2.0.CO%3B2).

77. NASA Earth Observatory, (June 14, 2009), *Climate and Earth's Energy Budget*, (https://earthobservatory.nasa.gov/features/EnergyBalance/page6.php).

78. Watts, Anthony, (May 2009), *Is the U.S. Surface Temperature Record Reliable?*, Watts Up With That? (http://static.cbslocal.com/station/wbz/wbz/2009/may/SurfaceStations.pdf).

79. The Physical Basis, Section 9.2.2.1, (2007), *Spatial and Temporal Patterns of Response*, IPCC-AR4 Working Group I, (http://www.ipcc.ch/publications_and_data/ar4/wg1/en/ch9s9-2-2.html).

80. Evans, Dr. David, (June 12, 2016), *The Missing Greenhouse Signature*, Principia Scientific International, (https://principia-scientific.org/the-missing-greenhouse-signature/).

81. Wallace III, Dr. James P.; Christy, Dr. John R. and D'Aleo, Joseph S., (September 2016), *On the Existence of a "Tropical Hot Spot" & The Validity of EPA's CO$_2$ Endangerment Finding*—Research Report Executive Summary, (https://thsresearch.files.wordpress.com/2016/09/ef-cpp-sc-2016-data-ths-paper-jpw-exsum-0915161.pdf).

82. Wallace III, Dr. James P.; Christy, Dr. John R. and D'Aleo, Joseph S., (September 17, 2016), *On the Existence of a "Tropical Hot Spot" & The Validity of EPA's CO$_2$ Endangerment Finding*—Abridged Research Report, (https://thsresearch.files.wordpress.com/2016/09/wwww-ths-rr-091716.pdf).

83. Sherwood, Steve, (June 3, 2015), *Climate meme debunked as the 'tropospheric hot spot' is found*, (https://theconversation.com/climate-meme-debunked-as-the-tropospheric-hot-spot-is-found-42055).

84. Bastasch, Michael (September 22, 2016), *Study: Tropical Hotspot 'Fingerprint' Of Global Warming Doesn't Exist In The Real World Data*, (https://wattsupwiththat.com/2016/09/22/study-tropical-hotspot-fingerprint-of-global-warming-doesnt-exist-in-the-real-world-data/).

85. Michaels, Pat and Knappenberger, Chip, (December 2015), *Climate models versus climate reality*, judithcurry.com, (https://judithcurry.com/2015/12/17/climate-models-versus-climate-reality/).

86. Ole Humlum website, *Climate4You*, (http://www.climate4you.com/).

87. Hezel, Paul J, (2013), *IPCC AR5 WGI: Polar Regions Polar Amplification, Permafrost, Sea ice changes*, Working Group I contribution to the IPCC Fifth Assessment Report, (https://unfccc.int/files/science/workstreams/research/application/pdf/5_wgiar5_hezel_sbsta40_short.pdf).

88. Richard, Kenneth, (April 16, 2018), *In 2015, Climate Scientists Wrecked Their Own CO$_2$-Forced 'Polar Amplification' Narrative*, NoTricksZone website, (http://notrickszone.com/2018/04/16/in-2015-climate-scientists-wrecked-their-own-CO2-forced-polar-amplification-narrative/).

89. Schmithüsen, Holger, et al, (November 25, 2015), *How increasing CO$_2$ leads to an increased negative greenhouse effect in Antarctica*, Geophysical Research Letters (AGU Journal), (https://agupubs.onlinelibrary.wiley.com/doi/full/10.1002/2015GL066749).

90. Nova, Joanne, (April 2018), *Climate change means Greenland is the same temperature now as 1880*, JoanneNova.com.au, (http://joannenova.com.au/2018/04/greenland-same-temperature-now-as-1880/).

91. Bastasch, Michael, (July 2017), *Bombshell study: Temperature Adjustments Account For "Nearly All Of The Warming" In Government Climate Data*, Watts Up With That? (https://wattsupwiththat.com/2017/07/06/bombshell-study-temperature-adjustments-account-for-nearly-all-of-the-warming-in-government-climate-data/).

92. Clutz, Ron, (July 2017), *Man Made Warming from Adjusting Data*, Science Matters, (https://rclutz.wordpress.com/2017/07/27/man-made-warming-from-adjusting-data/).

93. Bell, Larry, (October 2013), *IPCC In A Stew: How They Cooked Their Latest Climate Books*, Forbes (https://www.forbes.com/sites/larrybell/2013/10/13/ipcc-in-a-stew-how-they-cooked-their-latest-climate-books/#33c32ff12edd).

94. Godwin, Brendan, (December 2016), *Homogenization of Temperature Data By the Bureau of Meteorology* (Australia), (https://wattsupwiththat.com/2016/12/21/homogenization-of-temperature-data-by-the-bureau-of-meteorology/).

95. Delingpole, James, (November 24, 2015), *German Professor: NASA Has Fiddled Climate Data On "Unbelievable" Scale*, Breitbart, (https://www.breitbart.com/politics/2015/11/24/german-professor-nasa-fiddled-climate-data-unbelievable-scale/).

96. Editorial, (April 11, 2018), *A Startling New Discovery Could Destroy All Those Global Warming Doomsday Forecasts*, Investor's Business Daily, (https://www.investors.com/politics/editorials/global-warming-computer-model-nitrogen-rocks/).

97. Heller, Tony, (June 23, 2014), *NOAA/NASA Dramatically Altered US Temperatures After The Year 2000*, "Real Science" Climate Science Blog, (https://stevengoddard.wordpress.com/2014/06/23/noaanasa-dramatically-altered-us-temperatures-after-the-year-2000/).

98. Orwell, George, (1949), *Nineteen Eighty-Four*, Winston Smith, a character, (https://en.wikipedia.org/wiki/Winston_Smith).

99. Delingpole, James (February 20, 2018), *NOAA Caught Adjusting Big Freeze out of Existence*, Breitbart, (http://www.breitbart.com/politics/2018/02/20/delingpole-noaa-caught-adjusting-big-freeze-out-of-existence/?mc_cid=4d0cb72fa1&mc_eid=73a7cfe547).

100. Homewood, Paul, (January 25, 2018), *New York's Temperature Record Massively Altered By NOAA*, Not A Lot of People Know That (website), (https://notalotofpeopleknowthat.wordpress.com/2018/01/25/new-yorks-temperature-record-massively-altered-by-noaa/).

101. Heller, Tony, (August 23, 2017), *100% Of US Warming Is Fake*, The Deplorable Climate Science Blog (realclimatescience.com), (https://realclimatescience.com/2017/08/100-of-us-warming-is-fake/).

102. Menton, Francis, January 4, 2017, *The Greatest Scientific Fraud Of All Time—Part XI,* Manhattan Contrarian, (http://manhattancontrarian.com/blog/2017/1/4/the-greatest-scientific-fraud-of-all-time-part-xi).

103. Orwell, George (1949), *1984*, Secker & Warburg, (https://www.cliffsnotes.com/literature/n/1984/book-summary and https://en.wikipedia.org/wiki/Nineteen_Eighty-Four).

104. Solomon, Lawrence (2010), *The Deniers, Fully Revised: The World-Renowned Scientists Who Stood Up Against Global Warming Hysteria, Political Persecution and Fraud*, Richard Vigilante Books, (https://www.amazon.com/Deniers-Fully-Revised-World-Renowned-Persecution/dp/0980076374).

105. Homewood, Paul (May 10, 2014), *The Heat Waves Of The 1930's,* Not A Lot Of People Know That, (https://notalotofpeopleknowthat.wordpress.com/2014/05/10/the-heat-waves-of-the-1930s/).

106. Heller, Tony, (April, 28, 2018), *Plummeting Summer Temperatures In The US*, realclimatescience.com, (https://realclimatescience.com/2018/04/plummeting-summer-temperatures-in-the-us/).

107. NOAA, (May 8, 2018), *U.S. had its coldest April in more than 20 years*, (http://www.noaa.gov/news/us-had-its-coldest-april-in-more-than-20-years).

108. Gosselin, P. (November 11, 2018), *Arctic Sea Ice Soars, Polar Bears Start Hunt Early – 2nd Year In A Row!*, NoTricksZone.com, (http://notrickszone.com/2018/11/11/arctic-sea-ice-soars-polar-bears-start-hunt-early-2nd-year-in-a-row/).

109. Wikipedia, *U.S. state temperature extremes*, (https://en.wikipedia.org/wiki/U.S._state_temperature_extremes).

110. Switalski, Bernard, (April 3, 2008), *Environmentalism's Tainted Roots*, (http://www.webcommentary.com/php/ShowArticle.php?id=switabern&date=080403).

111. Watts, Anthony, (March 1, 2013), *The 1970's Global Cooling Compilation – looks much like today*, Watts Up With That? (https://wattsupwiththat.com/2013/03/01/global-cooling-compilation/).

112. Jackson, Kerry, (February 2, 2017), *If Global Warming Is Real, Why Do Government Scientists Have To Keep Cheating?*, Investor's Business Daily, (https://www.investors.com/politics/commentary/if-global-warming-is-real-why-do-government-scientists-have-to-keep-cheating/).

113. *The Solar Evidence*, (May 5, 2010), Global Warming Science, (http://appinsys.com/GlobalWarming/GW_Part6_SolarEvidence.htm).

114. NASA, (March 20, 2003), *NASA Study Finds Increasing Solar Trend That Can Change Climate*, NASA GISS, (https://www.giss.nasa.gov/research/news/20030320/).

115. Soon, W.H., (2005), *Variable Solar Irradiance as a Plausible Agent for Multidecadal Variations in the Arctic-wide Surface Air Temperature Record of the Past 130 Years*, Geophysical Research Letters, Vol. 32, (http://www.agu.org/pubs/crossref/2005/2005GL023429.shtml).

116. LASP (U. of Colorado), (March 2018), *Total Solar Irradiance Data*, SORCE, Laboratory for Atmospheric and Space Physics, U. of Colorado, Boulder, (http://lasp.colorado.edu/home/sorce/data/tsi-data/).

117. Andersen, Morten G., (August 25, 2016), *Solar activity has a direct impact on Earth's cloud cover*, Technical University of Denmark, Phys. Org, (https://phys.org/news/2016-08-solar-impact-earth-cloud.html).

118. CDIAC (now, ESS-DIVE), *Frequently Asked Global Change Questions: What percentage of the CO_2 in the atmosphere has been produced by human beings through the burning of fossil fuels?* US Government's Carbon Dioxide Information Analysis Center, (http://cdiac.ornl.gov/pns/faq.html).

119. Christy, John (PhD), (March 29, 2017), *Testimony of John R. Christy to U.S. House Committee on Science, Space & Technology*, science.hous.gov, (https://science.house.gov/sites/republicans.science.house.gov/files/documents/HHRG-115-SY-WState-JChristy-20170329.pdf).

120. Usoskin, I.G., et al, (August 6, 2014), *A 3,000-Year Record of Solar Activity*, CO_2 Science, (http://www.CO2science.org/articles/V17/N32/C1.php).

121. UK Natural History Museum, *Piltdown Man*, (http://www.nhm.ac.uk/our-science/departments-and-staff/library-and-archives/collections/piltdown-man.html).

122. Sallée, Jean-Baptiste[a]; Matear, Richard J.[b]; Rintoul, Stephen R.[b,c]; Lenton, Andrew[b,] *Localised subduction of anthropogenic carbon dioxide in the Southern Hemisphere oceans*, [a]British Antarctic Survey, [b]CSIRO Marine and Atmospheric Research, Wealth from Oceans National Research Flagship, [c]Antarctic Climate and Ecosystems Cooperative Research Centre, (https://core.ac.uk/download/pdf/9697160.pdf).

123. Martin, (July 12, 2013), *The Stork-and-Baby Trap*, PsychologyToday.com, (https://www.psychologytoday.com/us/blog/how-we-do-it/201307/the-stork-and-baby-trap).

124. Wikipedia, *El Niño—Southern Oscillation*, Wikipedia, (https://en.wikipedia.org/wiki/El_Niño—Southern_Oscillation).

125. United Nations, (archive of January 8, 2014, retrieved May 7, 2018), *Introduction to the Convention*, United Nations Framework Convention on Climate Change (UNFCCC), (https://web.archive.org/web/20140108192827/http://unfccc.int/essential_background/convention/items/6036.php).

126. Wikipedia, *Isotope*, (https://en.wikipedia.org/wiki/Isotope).

127. Dictionary.com, *pH defined*, (http://www.dictionary.com/browse/ph).

128. The Grid, (September 17, 2015), *What is the difference between solar insolation and solar irradiance?*, The Grid [solar electric site], (https://thegrid.rexel.com/en-us/energy_efficiency/f/energy_efficiency_forum/476/what-is-the-difference-between-solar-insolation-and-solar-irradiance).

129. Nola Taylor Redd, (May 18, 2018), *What is Solar Wind?*, space.com, (https://www.space.com/22215-solar-wind.html).

130. NASA, (2018), *Solar Minimum is Coming*, NASA Science, (https://science.nasa.gov/science-news/news-articles/solar-minimum-is-coming).

131. Watts, Anthony, (August 6, 2014), *Recent paper finds 1950-2009 Solar Grand Maximum was a 'rare or even unique event' in 3,000 years*, Watts Up With That?, (https://wattsupwiththat.com/2014/08/06/recent-paper-finds-recent-solar-grand-maximum-was-a-rare-or-even-unique-event-in-3000-years/).

132. Usoskin, I.G., Hulot, G., Gallet, Y., Roth, R., Licht, A., Joos, F., Kovaltsov, G.A., Thebault, E. and Khokhlov, A., (August 2014), *Evidence for distinct modes of solar activity (A 3,000-Year Record of Solar Activity)*, Astronomy and Astrophysics 562: L10, (http://www.CO2science.org/articles/V17/N32/C1.php).

133. Rice, Doyle, (March 15, 2017), *Climate change is making us sick, top U.S. doctors say*, USA Today, (https://www.usatoday.com/story/news/health/2017/03/15/climate-change-making-us-sick-top-us-doctors-say/99218946/).

134. Rice, Doyle, (June 29, 2017), *Who will pay most for climate change? South will be biggest loser*, USA Today, (https://www.usatoday.com/story/news/nation/2017/06/29/economic-cost-climate-change-southern-u-s-biggest-loser/439362001/).

135. Bomey, Nathan, (June 1, 2017, updated), *How Trump's energy order in March affects jobs, fuel prices*, USA Today, (https://www.usatoday.com/story/money/2017/03/28/trump-energy-regulations-impact/99700640/).

136. Drash, Wayne, (September 19, 2017), *Yes, climate change made Harvey and Irma worse*, CNN, (https://www.cnn.com/2017/09/15/us/climate-change-hurricanes-harvey-and-irma/index.html).

137. Masters, Jeffrey (PhD), *A Detailed View of the Storm Surge: Comparing Katrina to Camille*, Weather Underground—wunderground.com, (https://www.wunderground.com/hurricane/surge_details.asp).

138. Rice, Doyle, (April 23, 2018), *California's wild extremes of drought and floods to worsen as climate warms*, USA Today, (https://www.usatoday.com/story/news/nation/2018/04/23/california-drought-floods-worsen-climate-change/541822002/).

139. Canlorbe, Grégoire, (October 28, 2017), *Interview with István Markó, for Breitbart News Network—unabridged version*, gregoirecanlorbe.com, (http://gregoirecanlorbe.com/interview-with-istvan-marko-for-breitbart-news-network).

140. Rice, Doyle, (May 14, 2018), *Supercharged by global warming, record hot seawater fueled Hurricane Harvey* (updated May 15), USA Today, (https://www.usatoday.com/story/news/2018/05/14/hurricane-harvey-record-hot-seawater-global-warming/607715002/).

141. Ingham, John, (April 24, 2018), *Climate change is 'not as bad as we thought' say scientists*, express.co.uk (Daily & Sunday Express), (https://www.express.co.uk/news/uk/950748/climate-change-scientists-impact-not-as-bad-on-planet).

142. Segalstad, T.V. (August 5, 2009), *Correct timing is everything—also for CO_2 in the air*, CO_2 Science, Vol. 12, No. 31, (http://www.CO2science.org/articles/V12/N31/EDIT.php) also (http://www.CO2web.info/Segalstad_CO2-Science_090805.pdf).

143. Dr. Tom V. Segalstad website, *CO_2 Web*, (http://www.CO2web.info).

144. Anthony Watts' website, *Watts Up With That?* (https://wattsupwiththat.com).

145. *Committee For A Constructive Tomorrow (CFACT)*, website, (http://www.cfact.org).

146. *Principia Scientific International*, (http://principia-scientific.org).

147. Hans Schreuder's website, *I Love My Carbon Dioxide* (http://www.ilovemycarbondioxide.com).

148. Francis Menton's website, *Manhattan Contrarian* (http://www.manhattancontrarian.com).

149. *Climate Change Dispatch*, (http://www.climatechangedispatch.com).

150. Mark Morano's CFACT website, *Climate Depot*, (http://www.climatedepot.com).

151. Joe D'Aleo's website, *ICECAP* (http://www.icecap.us).

152. Tony Heller's website, *The Deplorable Climate Science Blog*, (https://realclimatescience.com).

153. Dr. Judith Curry's website, (https://judithcurry.com).

154. Ron Clutz's website, *Science Matters* (https://rclutz.wordpress.com/2017/10/26/global-ocean-cooling-in-september/).

155. United Nations, (revised version of April 20, 2018), *The Paris Agreement, United Nations Framework Convention on Climate Change*, UNFCCC, (https://unfccc.int/process-and-meetings/the-paris-agreement/the-paris-agreement).

156. NOAA PMEL Carbon Program, *What is Ocean Acidification?*, (https://www.pmel.noaa.gov/CO2/story/What+is+Ocean+Acidification%3F).

157. Elmhurst College, (2003), *Acids and Bases*, (http://chemistry.elmhurst.edu/vchembook/184ph.html).

158. National Geographic, (April 27, 2017), *Ocean Acidification*, (https://www.nationalgeographic.com/environment/oceans/critical-issues-ocean-acidification/).

159. Delingpole, James, (April 30, 2016), *Ocean acidification: yet another wobbly pillar of climate alarmism*, The Spectator, (https://www.spectator.co.uk/2016/04/ocean-acidification-yet-another-wobbly-pillar-of-climate-alarmism/).

160. Wikipedia, (March 27, 2019), *Atmospheric methane*, (https://en.wikipedia.org/wiki/Atmospheric_methane).

161. Carr, Ada, (October 26, 2017), *Volcanic Eruptions May Be Rapidly Melting Arctic Ice Sheets, Study Says*, The Weather Channel, (https://weather.com/news/climate/news/2017-10-25-arctic-sea-ice-volcanic-eruption-trigger-melting).

162. Global Carbon Project (website), (2018), *Global Carbon Budget—Summary Highlights*, (http://www.globalcarbonproject.org/carbon-budget/17/highlights.htm).

163. Hollingsworth, Barbara, (September 14, 2016), *Four Studies Find "No Observable Sea-Level Effect" From Man-Made Global Warming*, (https://www.cnsnews.com/news/article/barbara-hollingsworth/4-peer-reviewed-studies-find-no-observable-sea-level-effect-man).

164. Fox, Porter, (February 7, 2014), *The End of Snow?*, (https://www.nytimes.com/2014/02/08/opinion/sunday/the-end-of-snow.html).

165. Bastasch, Michael, (March 4, 2014), *Top 5 failed 'snow free' and 'ice free' predictions*, The Daily Caller, (http://dailycaller.com/2014/03/04/top-5-failed-snow-free-and-ice-free-predictions/).

166. Erdman, Jon, (March 23, 2015), *New England Record Snow Tracker: Boston Breaks All Time Seasonal Snow Record in 2014-2015*, The Weather Channel, (https://weather.com/news/news/new-england-boston-record-snow-tracker).

167. Caribou, Maine, *A Record Setting Snowy 7 to 10 Days for Downeast Maine*, Weather Forecast Office, (http://www.weather.gov/car/recordsnowystretch).

168. Wikipedia, (January 2017), *January 2017 European cold wave*, (https://en.wikipedia.org/wiki/January_2017_European_cold_wave).

169. UK Independent, (December 21, 2016), *First Sahara desert snow in 40 years captured in photographs*, (http://www.independent.co.uk/news/world/africa/sahara-desert-snow-first-40-years-rare-photos-atlas-mountains-algeria-karim-bouchetata-a7488056.html).

170. The Washington Post, (December 26, 2017), *Unforgiving cold snap will engulf eastern two-thirds of the nation through New Year's Day*, (https://www.washingtonpost.com/news/capital-weather-gang/wp/2017/12/26/unforgiving-cold-snap-will-engulf-eastern-two-thirds-of-the-nation-through-new-years-day/?utm_term=.c452f10dbd6b).

171. NOAA, *Frequently Asked Question about Radiosonde Data Quality*, (https://www.weather.gov/upperair/FAQ-QC).

172. Olson, Walter, (October 1, 2015), *Should Climate Change Deniers Be Prosecuted?* Newsweek, (http://www.newsweek.com/should-climate-change-deniers-be-prosecuted-378652).

173. Greenfield, Daniel, (December 25, 2012), *Progressive Professor Demands Death Penalty for Global Warming*, Frontpage Magazine, (https://www.frontpagemag.com/point/170948/progressive-professor-demands-death-penalty-global-daniel-greenfield).

174. USCRN, NOAA, (September 2019), *Contiguous U.S. Average Temperature Anomaly (degrees F)*, (https://www.ncdc.noaa.gov/temp-and-precip/national-temperature-index/time-series.csv?parameter=anom-tavg&time_scale=p12&month=12&datasets%5B%5D=uscrn).

175. USCRN, NOAA, *U.S. Climate Reference Network*, (https://www.ncdc.noaa.gov/crn/).

176. IPCC, (2014), *Climate Change 2014: Synthesis Report, Contribution of Working Groups I, II, and III to the Fifth Assessment Report of the Intergovernmental Panel on Climate Change*, (https://ar5-syr.ipcc.ch/ipcc/ipcc/resources/pdf/IPCC_SynthesisReport.pdf).

177. Watts, Anthony, (July 30, 2019), "'Hidden' NOAA temperature data reveals that 6 of the last 9 months were below normal in the USA – and NOAA can't even get June right," *Watts Up With That?*, (https://wattsupwiththat.com/2019/07/30/hidden-noaa-temperature-data-reveals-that-6-of-the-last-9-months-were-below-normal-in-the-usa-and-noaa-cant-even-get-june-right/).

ACKNOWLEDGEMENTS

I am most grateful to Dr. Tom V. Segalstad and to Alan Siddons for their many communications, studies, articles, suggestions, and helpful sources of pertinent material. Dr. Segalstad's work with the lifetime of carbon dioxide emissions is a key element of the case exonerating fossil fuel use from any responsibility for the growth of atmospheric CO_2. It was at the suggestion of Dr. Segalstad that this book was written.

The insights, comments, and suggestions from Carl Brehmer; Greg Crook; Dr. Pierre R. Latour; Christopher Monckton, third Viscount Monckton of Brenchley; and the late Hans Schreuder have been particularly valuable and helpful and are very much appreciated.

ABOUT THE AUTHOR

Bob Webster (Robert Dartt Webster) graduated from Virginia Polytechnic Institute in 1964 with a BS degree in Mathematics. Having held positions as a Mathematician, Mathematician-Statistician, Physical Science Analyst, Operations Research Analyst, and Systems Analyst, he retired after more than thirty years employment with the US Army and US Navy within the Department of Defense.

Bob's lifelong fascination with science includes particular interests in meteorology, climatology, geology, and astronomy. At age fourteen, he subscribed to the US Weather Bureau's *Daily Weather Maps*, which he continued until publication ceased several decades later. Beginning with a period of global cooling that peaked between the mid-1950s and late 1970s, for more than sixty years, widespread concerns about climate change have been a key interest of his. Bob's years in public school and college were embedded within a four decade mid-20th-century cooling period that spawned fears among scientists that the planet might be experiencing the onset of a new ice age. The cooling climate made a particular impression upon him as colder temperatures, ice skating, and impressive fall, winter, and spring snowstorms fueled his interest in weather and climate. The sudden reversal of climate fears to those of global warming in the 1980s served to heighten Bob's interest in discovering the truth behind changing climate.

A strong reverence for truth and an appreciation for *the scientific method* prompted Bob to carefully scrutinize the US government's global atmospheric carbon dioxide and temperature records to discover whether or not the IPCC's climate change theory is supported by the record of real-world observations. That examination exposed stunning contradictions to climate change theory that made a compelling case to write this book.

Bob and his wife Joanne live in Indian River County, Florida.

CPSIA information can be obtained
at www.ICGtesting.com
Printed in the USA
BVHW051139030222
627688BV00001B/1

* 9 7 8 1 6 6 2 4 2 9 2 0 0 *